For the listeners of our show, who have kept us
inspired every day with their incredible curiosity

About This Book

The illustrations for this book were rendered digitally. This book was edited by Samantha Gentry
and designed by Neil Swaab. The production was supervised by Bernadette Flinn, and the
production editor was Jen Graham. The text was set in Bembo Std, and the display type is
KG Second Chances and KG Tangles Up In You 2.

Little, Brown and Company
Hachette Book Group
1290 Avenue of the Americas, New York, NY 10104
Visit us at LBYR.com

First Edition: September 2020

Little, Brown and Company is a division of Hachette Book Group, Inc.
The Little, Brown name and logo are trademarks of Hachette Book Group, Inc.

The publisher is not responsible for websites (or their content) that are not owned by the publisher.

Library of Congress Cataloging-in-Publication Data
Names: Bloom, Molly Hunegs, 1983– author.
Title: Brains on! presents...it's alive : from neurons and narwhals to the fungus among us /
Molly Bloom, Sanden Totten, and Marc Sanchez.
Description: First edition. | New York : Little, Brown and Company, 2020. | Includes bibliographical
references and index. | Audience: Ages 8–12 | Summary: "The team from the popular kids' science podcast
Brains On! brings readers a humorous, fact- and fun-filled look at all things biology—from animals and plants
to the human body and the microverse." —Provided by publisher.
Identifiers: LCCN 2019048842 | ISBN 9780316428293 | ISBN 9780316428316 (ebook) |
ISBN 9780316428309 (ebook other)
Subjects: LCSH: Biology—Juvenile literature.
Classification: LCC QH309.2 .B59 2020 | DDC 570—dc23
LC record available at https://lccn.loc.gov/2019048842

ISBNs: 978-0-316-42829-3 (hardcover), 978-0-316-42831-6 (ebook),
978-0-316-42871-2 (ebook), 978-0-316-42880-4 (ebook)

Printed in the United States of America

LSC-C

10 9 8 7 6 5 4 3 2 1

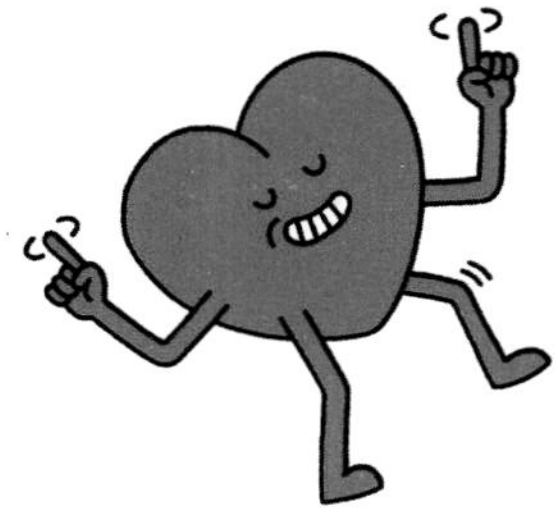

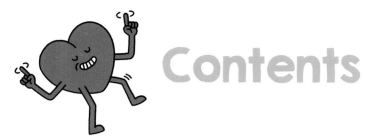

Contents

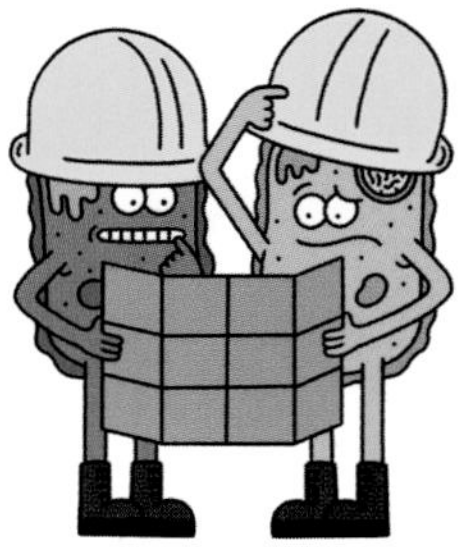

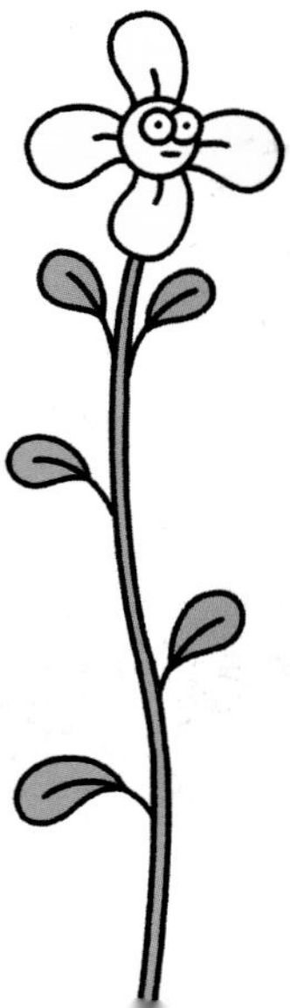

Introduction

We're the hosts of the podcast *Brains On!* We answer important questions like "Does the universe go on forever?" "What happens when we dream?" and "What are boogers anyway?"

If you're reading this book, then CONGRATULATIONS! You're alive, and you are about to have your mind blown! Unless you are a book-reading robot. In which case, you're not alive and you don't have a mind, just a bunch of computer chips and wires and stuff. Sorry, pal!

This book is about the totally bizarre, ultra-epic, sometimes disgusting world of biology. That's the study of living things, from minuscule bacteria and teeny-tiny ants to colossal trees and ginormous whales. We're going to explore this world and meet some pretty cool life-forms along the way, like superpowered pets, shape-shifting sea creatures, plants that bite back, brains that run on electricity, and micro mites living on your face! There will also be roller-coaster rides, shrink rays, laser-eyed cats, talking flowers, professional fart makers, and a whole lot more.

So when you start to read, be ready. You're about to get a megadose of super cool science that will fill up your brain and change how you see the world forever. You won't just learn about biology…you'll watch it come ALIVE!

PART I
ANIMALS

THE SECRET LIVES OF DOGS AND CATS

DO DOGS KNOW THEY'RE DOGS?

Dogs are always by our side, kind of like a shadow…if shadows pooped and hated squirrels. And we love them so much they are practically family. They comfort us when we're sad, slobber us with kisses when we come home, and take their job as head of security VERY seriously. Just ask the mail carrier.

At the same time, dogs love eating trash, burying bones, and sniffing other dogs' butts. Stuff we'd never do. Which brings up an important question: Do dogs think they're human, or do they know they're dogs?

Oh, Say, Can You See?

Dogs don't talk. So we can't ask them, "Hey, do you know you're a dog?" But there is an experiment that scientists use to see if animals are self-aware. In other words, do they know they are unique from all other beings? This experiment is called the mirror test.

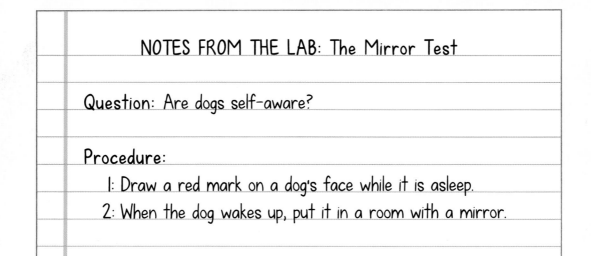

NOTES FROM THE LAB: The Mirror Test

Question: Are dogs self-aware?

Procedure:

   1: Draw a red mark on a dog's face while it is asleep.

   2: When the dog wakes up, put it in a room with a mirror.

> **3:** Wait to see if the dog looks in the mirror, notices the dot, and tries to wipe it off or inspect it somehow.
>
> **Results:** Dogs that were part of the test did not try to remove the dot. In fact, they might have thought the animal in the mirror was a different dog altogether!
>
> **Conclusion:** Dogs are not self-aware—at least not when they look in a mirror.

So dogs fail the mirror test. But that doesn't mean they aren't self-aware. Maybe all they need is a different test. You see, humans tend to rely on their eyes to sense the world around them. But dogs use their sensitive snouts to *smell* the world. In fact, their eyesight is not very good. But their sense of smell is so much better than we can even imagine. This might be why dogs don't recognize themselves in the mirror—the mirror dog has no smell.

ANIMALS THAT PASS THE MIRROR TEST

- Dolphins
- Elephants
- Chimpanzees
- Humans (but not babies)

SUPER COOL SCIENTIST
DR. ALEXANDRA HOROWITZ

Dr. Alexandra Horowitz runs the Dog Cognition Lab at Barnard College, Columbia University, and wants to know what it's like to be a dog. She studies how dogs think, see, and smell, and she developed a scent test to see if dogs are self-aware. Alexandra hopes her work is helping people better understand dogs. While we love dogs, there's a lot we don't know about them, and we wind up thinking they are misbehaving when they're really just trying to fit into a human world.

NOTES FROM THE LAB: The Scent Test

Question: Are dogs aware of their own smell?

Procedure:
  1: Take a little bit of a dog's urine and mix it with the urine of another dog. For a dog, this would be like seeing a picture of yourself mixed with a picture of someone else. Freaky!
  2: Leave samples of the dog's urine and the mixed version where it will find them both.
  3: See if the dog recognizes its own urine and expresses interest in the mixed version by sniffing it longer.

Results: The dogs were way more curious about the mixed version of their scent than the original. Just as humans use their eyes to see if their face has a dot on it, dogs use their nose to realize their scent has changed. Alexandra Horowitz, who developed this experiment, calls it an olfactory mirror test.

Conclusion: Dogs do seem to recognize themselves, but it's rooted in their sense of smell, not their sense of sight.

This pee smells weird, right? Take a whiff.

no, thanks. I trust you.

Olfactory: of or relating to the sense of smell

The Way to a Dog's Mind Is Through Its Nose!

Smells might even be a way for dogs to tell time. Dr. Alexandra Horowitz thinks dogs might know how long you've been away from home just by your lingering scent. When you leave for school in the morning, you leave your scent behind. As the day goes on, your scent gets weaker and begins to disappear. Dogs, with their super-smelling machines, might know that when your scent is at half its normal strength, you usually come home from school. So when that happens, it's time to wait by the door and wag that tail!

Dogs can remember you after you've been gone for a day, a week, a month, or even a year, because they remember your scent. That's why the golden retriever you see once a year at Thanksgiving knows you're the one who will slip it a piece of turkey under the table.

It's impossible to know if dogs remember specific events in their lives—at least until we invent a bark translator. If a dog had a scary

My clock says it's time for belly rubs.

Again?

This bark translator prototype needs some work. Every bark reads as just "peanut butter!"

THE INTERNATIONAL LANGUAGE OF BARKS

Dogs speak the same way no matter where they're from, but did you know that humans hear barks differently depending on the language they speak?

WHEN OUR DOGS HEAR US PUT THE KEY IN THE DOOR, THEY GO BONKERS AND SAY...

English	Woof, woof
Spanish	Guau, guau
Somali	Wuh, wuh
Hmong	Bow, vow, vow
Russian	Guff, guff
Polish	How, how
Mandarin Chinese	Wang, wang

experience at a shelter as a puppy or at the vet's office, it might not remember *exactly* what happened. But it might remember enough about the smells to be scared if you take it back.

SWEET DOGGY DREAMS

Have you ever seen a dog run in place while it sleeps? Or growl at something during naptime? Chances are you've caught your dog dreaming. Dogs spend about half of their day sleeping, but they dream for only a small amount of that time. That's probably because they take lots of short naps instead of one long slumber like us.

Dogs have both rapid eye movement (REM) sleep and non-rapid eye movement sleep just like humans. Dogs enter REM about twenty minutes into a snooze. And REM sleep is where a dog's fantasies finally take flight!

Sure, dogs twitch and growl, but luckily they don't get up and chase dream squirrels while snoozing. That's thanks to a part of the brain called the pons. It keeps dogs, humans, and other animals from acting out their dreams. In puppies, the pons isn't yet fully developed, and in older dogs, it can be less efficient. In those cases, a dozing dog might get up and sleepwalk. Watch out, dream squirrels…and any tables or lamps in the way!

Without a pons, a sleeping pointer might point its nose at dream birds, or a terrier might dig for dream rabbits. My dog would probably eat dream trash.

(For more about human dreams, see page 101.)

BACK TO THE FUTURE

Can dogs time-travel? Well, kind of. Dogs know what happened in the past by scent tracks, which means you can't keep any secrets from your dog. Imagine if you took a detour on your way home from school. You went on a quick mountain bike ride with your friend. Then, because you were *starving* from all that exercise, you grabbed a slice of pizza. When you stroll through the front door,

WHAT HAPPENS WHEN A DOG ENTERS DREAMLAND?

- Legs start to twitch
- Breathing gets shallower
- Eyes move rapidly behind their eyelids

your dog might take a few sniffs and think: "Hey, you hit the trails without me! And you went to the pizza place. You could have at least brought me home a slice." All this is based on your scent.

Dogs can also read the future…sort of. Their noses know what's blowing in the wind, so they can smell what's coming long before we can. If you're on a walk, your dog might sniff the breeze and know that there's a cat up ahead or that your neighbor just grilled a burger. They might even start wagging their tail because they know your best friend is right around the corner, even though you can't see them yet. Don't freak out. Your pup isn't psychic, just a really good sniffer.

WHAT'S THE SECRET OF DOGS' AMAZING SENSES?

Dogs can't read or write or play Mozart on the piano (though we'd totally pay to see that), but they *can* sense things we never could. Dogs have the same five senses as humans, but their sense of smell is about 10,000 times better (and their ears can hear far beyond our range; see page 11). Their pooch powers would make any superhero jealous.

Next-Level Noses

Dogs' noses aren't just the cutest, they're also amazing sniffing machines. The moist, spongy outside helps them capture scents. Plus, they have a lot more olfactory receptors than humans do. And the part of a dog's brain that is devoted to analyzing smells is about forty times bigger than ours.

> For comparison, forty times bigger is like the difference between a golf ball and a honeydew melon.

Those olfactory receptors are great at detecting smells. It happens like this: Smells come in the form of tiny molecules floating in the air—bits of flowers in your yard, the chocolate chip cookie you ate after lunch, or your neighbor's overflowing garbage can. SNIFF! A dog's scent receptors grab on to those molecules and check them out. Then they send a message to the dog's brain letting it know that it's smelling a salami sandwich or another dog's butt.

> I think I smell a squirrel somewhere in this book! Page 38 to be exact!

But what does it mean when we say that dogs' sense of smell is so much better than humans'? If someone brought you a cup of water with a teaspoon of sugar in it, you *might* be able to smell that, but you'd have to taste it to be sure. A dog's nose, on the other hand, could detect a tiny teaspoon of sugar even if it was poured into a giant swimming pool.

Dogs can also smell with each nostril independently. If those sugar molecules enter the right side of the dog's nose and not the left, the dog will know that the sugar was poured into the right side of the pool.

Where humans would just smell chlorine at the pool, the dog would smell chlorine, that teeny bit of sugar, sunscreen, bathing suits, snorkels, goggles, and maybe even a little pee.

WHY ARE DOGS' NOSES WET?

When our noses are wet, that usually means we're sick. But dogs' noses are supposed to be that way! Dogs have sweat glands in their noses that help them keep their bodies at the right temperature. Their noses also secrete a thin layer of mucus that helps them capture scent molecules. And finally, dogs lick their noses a lot, both to clean them and to carry some of those scent particles to the Jacobson's organ for closer examination.

I Smell What You're Feeling

What does happy smell like? Or scared? These might sound like silly questions to us, but to dogs they make perfect sense. Dogs have a powerful smelling tool called Jacobson's organ. If you want to get really fancy, you can call it the vomeronasal organ. This bit of nose technology lets dogs smell moods by picking up chemicals called pheromones on another animal's body.

PRONOUNCER
Vomeronasal = vuh-MARE-oh-NAY-zull

If noses were vehicles, dogs would have a race car. Humans would have a tricycle...with a broken wheel.

Pheromones are chemicals that animals produce to send signals to one another. Like, if we humans sniffed a dog, we'd say, "Yep, that's definitely a dog." But if your dog Cuddles McMuffinsteen sniffed your neighbor's dog Penelope A. Poodle, her pheromones might tell Cuddles that Penelope was alarmed, angry, or looking for love.

To dogs, pheromones are like smellable mood rings!

I am trying to sniff how you're feeling, Sanden.

Good thing I'm wearing deodorant today...

ANIMALS WITH JACOBSON'S ORGAN

Humans don't have a working Jacobson's organ, but many other animals do, including:

- Hamsters
- Snakes
- Lizards
- Mice
- Rats
- Elephants
- Cats
- Goats
- Pigs
- Giraffes
- Horses
- Bears

MYSTERY PHOTO

Focus your eyes on this mystery photo. Can you guess what it is? Turn to the next page for the answer.

Gee, Your Butt Smells Friendly

You've probably seen dogs sniffing each other's rear ends. It's not because dogs are gross. Well, it *is* because they're a little gross, but it's also because there are glands back there that produce those important chemical signals—pheromones. So when a dog sniffs a butt, it's almost like looking at someone's social media. They can tell whether that dog is male or female, roughly how old it is, whether it's sick or healthy, and what kind of mood it might be in. Dogs can then use that information to start a new friendship or go in search of a friendlier butt.

DOGS TO THE RESCUE!

Dogs use their noses to do more than find out about the other hounds in town; they also use their scent receptors to help humans. They search for missing people in the wilderness, sniff out bombs in airports, and can even smell some types of cancer.

Some very special dogs are trained to help guide people who are visually impaired. Others can help people who are deaf or hard of hearing by alerting them when a doorbell rings or a smoke alarm goes off. Mobility assistance dogs help people who have trouble moving around. These pups can pick up dropped items, help open doors, or bring their owners the phone when it starts ringing! Now *that's* a good boy!

Ear-mazing!

Smell isn't the only super sense in dogs. They've got impressive ears too. Some dog breeds have better hearing than others, but they can all hear sounds that the human ear would never notice.

Dogs can hear higher-pitched sounds as well as those that are too quiet for human ears. That super hearing was once a matter of survival. Dogs evolved  from wolves, which hunt for small rodents like mice. So they need to hear those tiny, high-pitched squeaks to catch dinner and stay alive.

Dogs can also rotate, tilt, raise, and lower their ears to detect exactly where a sound is coming from. That's why your dog somehow always knows when you open a bag of chips!

Outta Sight!

Vision is one area where humans have dogs beat. A dog's vision is much blurrier than ours, and they also see fewer colors. Humans and dogs both have cells in their eyes that detect visible light—these are called cone cells. We have three kinds: cones that detect blue light, cones that detect green, and cones that detect red. These combine to let us see a wide range of colors. But dogs have only two kinds of cone cells: yellow and blue. They'd like to have a third cone: an ice cream cone! But that's not gonna happen.

WHY DO CATS' EYES GLOW IN THE DARK?

While it's fairly easy to guess a dog's emotions, cats are a bit more mysterious. The way they look at you, the way they sound, and even the way they raise their tails can have different meanings. If you want to understand the mind of a cat, you can invent a machine to switch minds and then live life as a feline. Or you can just read this next section. Honestly, that is much easier, with less chance of hair balls.

How Do Cats See in the Dark?

You know how your cat will choose the middle of the night to start its exercise routine? Dashing around the living room, jumping off couches, darting under tables—but somehow never running into the wall even though it's pitch-black? That's because cats need only a tiny sliver of light to navigate at night.

Rods and Cones

Rods and cones are the cells in our eyes that help us see. Rods are used to see in low light, and cones are used to detect color. These cells absorb light and send signals to the brain to say, "You're looking at a tree," or "Hey, there's something moving in the grass over there!" A typical human eye has about 120 million rods. Cats have *seven times* more than that, meaning their eyes are seven times more sensitive to light than ours. But rods don't pick up on color or fine details, so even good night vision is still pretty blurry.

Glowing Eyes

Cats, and many other animals, have a special reflective layer behind their retinas called the tapetum lucidum. It's sort of like a mirror in the back of the eye. It reflects incoming light so those light-detecting cells get a second chance to absorb it. If you've ever taken a picture of a cat at night and used the flash, you'll recognize those

MORE ANIMALS WITH EYES THAT GLOW IN THE DARK

- Deer
- Dogs
- Cows
- Horses
- Ferrets

eerie, glowing eyes. That's the tapetum lucidum reflecting the camera's flash back like a mirror. But it's also fun to pretend that it's because your pet has laser eyes. *Pew pew!*

Pupils

Check out your eyes in the mirror. Your pupils, the dark circles right in the middle of your eyes, are round. Cats' pupils, on the other hand, are the shape of a football standing on its end. These are called vertical pupils because they go from top to bottom. This shape lets cats adjust their pupils fast so they can open them very wide at night and let in lots of light. Human eyes are capable of greatness, but cats' eyes beat us hands down when it comes to night vision.

MYSTERY PHOTO

Focus your eyes on this mystery photo. Can you guess what it is? Turn to page 15 for the answer.

WHAT IS MY CAT TRYING TO TELL ME?

Sadly, Baron Fluffy von Whiskerface can't tell you what he wants with words, but if you read the clues, you might be able to decipher your cat's secret messages.

Eyes—Read Me If You Can

Cats use their eyes to communicate. When a cat's eyes are open very wide and its pupils are dilated, it's trying to take in more information from the world around it. It might be feeling playful or thinking about hunting. But here's the tricky part: A cat that is afraid will also have dilated pupils. So you need to look for more clues to find out how your cat is *really* feeling.

Puffy Tails—Back Off, Buddy

Ever wondered why your cat's tail puffs up when it sees another animal? Some cats even puff up all over. This is called piloerection. It's your cat's way of trying to look bigger and scarier than it really is. In other words, a big, puffed-up cat means run away stat!

MOMENT OF EW

CAT AND DOG MOMS EAT THEIR BABIES' POOP!

Be glad you aren't a cat or dog mom. For them, eating baby poop is the most natural thing in the world. Not only are the moms keeping their dens clean, but they're also protecting their newborn kittens or puppies from predators that are attracted to the smell. They do this from the time their kittens or puppies are born until they're big enough to leave the den to poop. Those moms should get the biggest, best Mother's Day card ever for doing that gross job.

ANSWER!

It's a cat's tongue! See those spikes covering the tongue? Those are called papillae, and they're made out of the same stuff as our fingernails. Cats use their tongues for grooming themselves, and these tiny claws are ideal for brushing through any tangles. This licking habit also helps them stay at a comfortable temperature. It distributes both protective oils, which can act as an insulator, and saliva, which cools them when it evaporates.

KITTY COMMUNICATION DECODED

Can you tell when your cat is hungry or wants a cuddle? Many scientists have studied how cats communicate with us, and they've discovered that cats have three main types of communication.

MEOWING: "HEY, PAY ATTENTION TO ME!"

- Kittens meow when they want attention from their mothers.

- Kittens and cats meow to humans when they want food or attention.

- Cats learn what humans respond to and tailor their meows to those particular humans. In fact, adult cats meow only around humans. They don't meow at other cats!

HISSING AND SCREECHING: "YOU WANT A PIECE OF ME?"

- Cats have a number of defensive, "don't mess with me" sounds, from hisses to growls.

- Loud sounds make them appear bigger and stronger and may help them avoid a fight.

- Cats sometimes scream at night when they're looking for a mate.

PURRING: "PAY EVEN MORE ATTENTION TO ME!"

- Some purrs come from happiness and contentment.

- Another purr is the "solicitation purr." This includes a higher-pitched tone than regular purrs, which makes it harder for humans to ignore. You hear this sound when your cat is hungry.

- Then there's the "I'm in pain" purr. Some scientists think that purring can help heal an injured cat.

The Hibbing Humans VS. The Akron Animals

MEGA MATCHUP
DOGS VS. CATS

Now it's time for a fierce and furry fight between our favorite domesticated beasts. In one corner, dogs! Those lickers of faces and chewers of toys! And in the other corner, cats! Those graceful mouse hunters with cute faces to boot. Which cuddly contender will come out on top?

TEAM DOG

- Dogs love you unconditionally forever. One dog in Japan named Hachiko waited for his owner every day at the train station, even for years after his owner passed away. Now a statue is there in his honor. That's maximum loyalty.

- Dogs save lives! They sniff out drugs and bombs, rescue people from dangerous places, and, in the case of Seeing Eye dogs, help people navigate the world safely.

- Kids who grow up with dogs in the house are less likely to get certain diseases. The theory is that dogs expose us to so many germs that our bodies become better at fighting off sickness. Dogs to the rescue—again!

- Dogs' noses are among the most amazing smelling machines on the planet. They have two sets of nostrils and more than 100 million scent receptors.

- Dogs have super hearing, and they can rotate, tilt, raise, and lower their ears to detect *exactly* where a sound is coming from.

TEAM CAT

- Cats are natural entertainers. That's why funny cat videos are all over the Internet.

- Cats are fiercely independent and can survive on their own. They don't really need us, but they probably enjoy the free food and shelter. Some might say they've trained us more than we've trained them.

- One of the biggest mysteries left in the world is the cat's purr. Why do they do it? Could it be a secret cat code? We might never know. But doesn't it sound soothing?

- Every part of a cat's eye is designed to give it super-powered night vision. From their millions of light-receptor cells and reflective layer to their vertical pupils, cats are the ultimate night hunters.

- Cats have sensitive whiskers that help them get around in the dark. When those whiskers brush up against things, they send signals letting the cat know what's around it. Whiskers: stylish *and* useful.

Which furry friend is cooler: Dogs or Cats?

YOUR VERDICT

CREATURES OF THE DEEP

HOW DO SEA CREATURES BREATHE UNDERWATER?

Fish can't play guitar, make a pizza, or even walk on land. But they *can* do something pretty awesome: They can breathe underwater. So can shrimp, squid, and more. There's plenty of oxygen dissolved in the oceans, but how do these creatures get it out of the water and into their bodies? That depends a lot on the animal. Some go above water to get air, and they hold their breath for a long time while swimming. Others breathe through gills. Jellyfish, for example, can absorb the oxygen in water directly through their skin. That's right—through their skin!

Gill Breathers

Fish are famous for their gills. But they aren't the only creatures that have them. Let's meet some other members of the

GILLionaires club!

- **Mollusks:** clams and oysters

- **Arthropods:** lobsters and crabs

- **Echinoderms:** starfish and sea urchins

- **Cephalopods:** cuttlefish and octopuses

Gills work pretty much the same way for all creatures that have them. Water moves over them, and inside are tiny fern-looking things, called filaments, that absorb oxygen from the water. The filaments move the oxygen into the bloodstream through diffusion.

Diffusion: The process of a substance spreading out to evenly fill its environment. In this case, oxygen moves from higher-oxygen areas to lower ones.

Lung Breathers

You might have noticed that some of your favorite sea creatures were missing from the list of gill breathers.

I once held my breath for thirty seconds. I probably could have gone longer, but there was a delicious plate of tacos on the table, and let's just say...you have to breathe when you eat, right?
#TeamTaco

- **Pinnipeds:** seals, sea lions, and walruses

- **Cetaceans:** whales, dolphins, and porpoises

- **Marine Fissipeds:** polar bears and sea otters

- **Sirenians:** manatees and dugongs

Mammals that live in the ocean still breathe air through lungs as we do. Most lung breathers take in air through nostrils, but cetaceans breathe through blowholes right on the top of their heads. After each breath, strong muscles around the blowhole seal it shut so that water can't get into the sea creature's lungs. Dolphins breathe about four or five times a minute but can hold their breath for up to fifteen minutes. Whales can go for longer, up to ninety minutes between breaths.

When whales dive deep, they make sure to pack a lot of oxygen. They carry it not just in their lungs but also in their blood and muscles. And they know how to make the most of the oxygen they've got. They can even slow down their heart rates to direct their oxygenated blood to their most important organs.

But even whales eventually have to surface for a fresh gulp of air. That's where blowholes come in. Scientists believe that long, long ago, the ancient relatives of whales had blowholes on their faces—or nostrils, as we know them. Over millions of years, those blowholes moved from their faces to the top of their heads. That's convenient for swimming and breathing at the same time. They have to put only the tops of their heads out of the water to exhale and inhale. Blowholes for the win!

Cell Breathers

Whales and fish are pretty rad, but wait until you hear about jellyfish. They don't have gills *or* lungs. They simply absorb the oxygen in water through the cells on their skin.

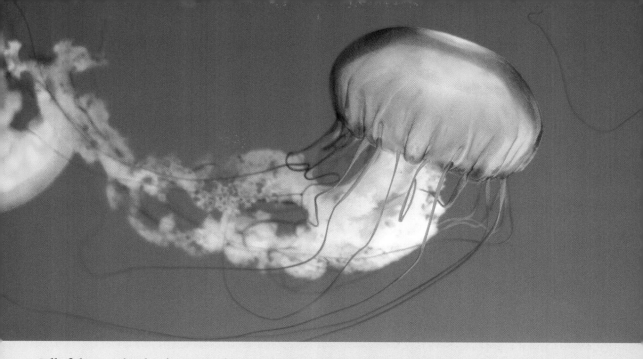

Jellyfish can do this because they're made up of just two cell layers—a thin layer on the surface of the body, and a thin layer lining the stomach. Sweet and simple. The rest of the creature is made of a type of jelly called mesoglea. Imagine where you would go if you could absorb oxygen through your skin.

I'd give an octopus eight high fives!

I'd take selfies with sea turtles.

I'd use seaweed to make a fabulous wig.

DEAD ZONES

Oxygen from the atmosphere dissolves in the water's surface and is spread by movement, like waves and currents. Areas of water that don't have enough oxygen to support life are called dead zones.

HOW DEAD ZONES FORM

1. Nutrients like nitrogen and phosphorus enter the water from farms and streets through runoff and storm drains.

2. Algae feed on these nutrients and grow out of control. The algae then sink and start to decompose.

3. Bacteria feasting on the dead algae use up all the oxygen in the water.

4. Fish can try to swim away if they can't get enough oxygen. But creatures like clams, oysters, mussels, and starfish will most likely die.

WHY DO NARWHALS HAVE HORNS?

Get ready to meet your new underwater BFF: the narwhal! Narwhals are a kind of whale. They're about as big as a car and have pale skin dappled with dark gray spots. If that wasn't lovely enough, they have one more distinctive feature: a long, spiraling "horn."

That's right, these bathing beauties are almost like IRL unicorns, but without the sparkles and rainbows or hooves and legs.

But really, it's not a horn at all. It's a tusk. A very long, spiral tooth that pokes out through a hole in their upper lip. It's the only spiral tooth in the natural world—at least that we know of. But what's the purpose of that weird and wonderful tusk-tooth?

Something to Sink Your Teeth Into

Narwhals aren't born with their tusks. In fact, most female narwhals

Fact-astic Voyage

Narwhals can live to be one hundred years old.

don't have tusks at all. As male narwhals mature, the tusk erupts out of the upper jaw and starts to grow. It can grow for up to twelve years and become almost as tall as a basketball hoop before the young male is an adult.

You'd think with that long tooth growing out of the front of its head, a narwhal would have a mouth full of teeth. But, in fact, narwhals don't have any teeth inside their mouths. They swallow their prey whole. They do have a second tooth that stays inside their skulls. In very rare cases, that second tooth grows, and you'll see a double-tusked narwhal.

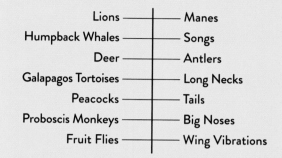

OTHER TRAITS TO ATTRACT A MATE

Lions	Manes
Humpback Whales	Songs
Deer	Antlers
Galapagos Tortoises	Long Necks
Peacocks	Tails
Proboscis Monkeys	Big Noses
Fruit Flies	Wing Vibrations

Scientists don't really know what the tusk's purpose is, but they think male narwhals use it most often to show off and attract females.

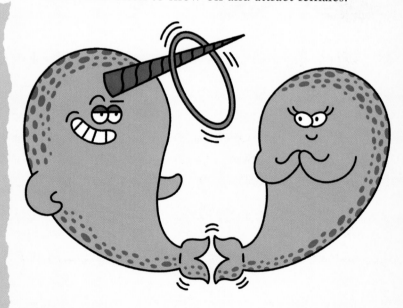

Historical Hoax

Hundreds of years ago, Vikings sailed the Arctic waters, traded goods with the Inuit peoples, and came home with narwhal tusks. But who would pay a lot of money for a whale's tooth? No one. Kings, however, were willing to fork over all kinds of treasure for unicorn horns. People believed they were magical. In fact, a Danish king had a throne built from unicorn horns. It wasn't until the late 1700s that the hoax was revealed: The chair was made of narwhal tusks.

But there may be other uses for that tusk-tastic tooth. Scientists have captured footage of male narwhals swimming through a school of fish and using their tusks to disturb the fish. Were the narwhals hunting? We aren't really sure. Scientists have had many hypotheses about the purpose of the tusk over the years—from an ice-breaking tool to a kind of weapon—but a lot more research is needed before humans will really know.

HOW DO WHALES MAKE SOUND?

There's no TV show called *The Ocean's Got Talent*, but there should be. It turns out that deep underwater, there are some amazing singers: whales! They whistle, squawk, and, yes, sing eerily beautiful songs. But exactly how do whales make those sounds?

My, What Strange Teeth You Have

There are two types of whales—baleen whales and toothed whales. Instead of teeth, baleen whales have something called baleen, which looks like a scrub brush crossed with a fine-tooth comb. It filters out water to retain the tiny ocean animals that these whales eat.

Some baleen whale species:

- Humpback Whales

- Right Whales

- Blue Whales

- Gray Whales

Did you know blue whales are the largest animals known to have ever lived on earth? A blue whale's tongue can weigh as much as a whole elephant. Their eyes are as big as grapefruits, and their hearts are the size of a Volkswagen Beetle.

MYSTERY PHOTO

Focus your eyes on this mystery photo. Can you guess what it is? Turn to page 28 for the answer.

Find the Right Bite at Crazy Eddies'

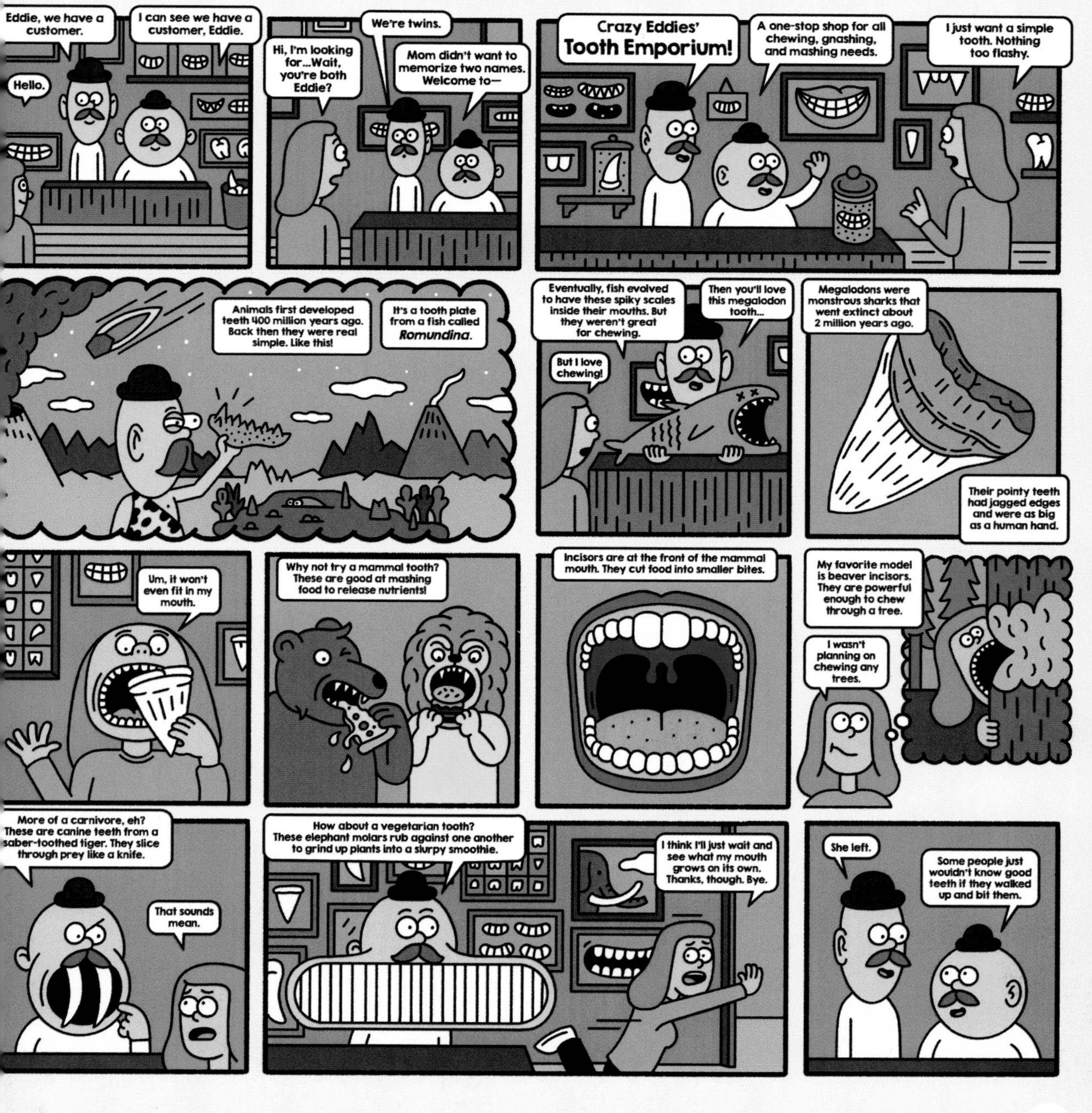
Eddie, we have a customer.
I can see we have a customer, Eddie.
Hello.
Hi, I'm looking for...Wait, you're both Eddie?
We're twins.
Mom didn't want to memorize two names. Welcome to—
Crazy Eddies' Tooth Emporium!
A one-stop shop for all chewing, gnashing, and mashing needs.
I just want a simple tooth. Nothing too flashy.
Animals first developed teeth 400 million years ago. Back then they were real simple. Like this!
It's a tooth plate from a fish called *Romundina*.
Eventually, fish evolved to have these spiky scales inside their mouths. But they weren't great for chewing.
But I love chewing!
Then you'll love this megalodon tooth...
Megalodons were monstrous sharks that went extinct about 2 million years ago.
Their pointy teeth had jagged edges and were as big as a human hand.
Um, it won't even fit in my mouth.
Why not try a mammal tooth? These are good at mashing food to release nutrients!
Incisors are at the front of the mammal mouth. They cut food into smaller bites.
My favorite model is beaver incisors. They are powerful enough to chew through a tree.
I wasn't planning on chewing any trees.
More of a carnivore, eh? These are canine teeth from a saber-toothed tiger. They slice through prey like a knife.
That sounds mean.
How about a vegetarian tooth? These elephant molars rub against one another to grind up plants into a slurpy smoothie.
I think I'll just wait and see what my mouth grows on its own. Thanks, though. Bye.
She left.
Some people just wouldn't know good teeth if they walked up and bit them.

Toothed whales catch their food, mostly squid and fish, by sucking them into their mouths or by chomping and biting.

Some toothed whale species:

- Orcas

- Narwhals

- Sperm Whales

- Belugas

Fact-astic Voyage

Dolphins are a kind of whale!

Even though whales don't have vocal cords, the way they make sound is similar to that of humans. When we talk, air moves over vocal cords in our throat. These cords then vibrate, making noise—kind of like the reed in a saxophone or clarinet.

Baleen whales don't have vocal cords, but they do have something called a U-fold at the back of their mouths. The U-fold is attached to a sac that moves air over the fold, making sound.

Toothed whales don't make sound through their mouths. They use their blowholes. Air moves through flaps called phonic lips and through their blowholes to make clicks and squawks.

My nose can't talk, but it can run! Get it? Anyone?

So a blowhole isn't just a kind of nose. It's a nose that talks!

ANSWER!

It's a group of krill! Baleen whales depend on krill to survive. These tiny members of the food chain are about the size of a paper clip. Whales open their mouths underwater to swallow as much krill as possible. Then they push the water back out through their baleen, leaving the krill behind in their mouth. So if you see a baleen whale, you can say, "Hey, you've got some krill in your grill!"

What's Up with Whale Talk?

Whale sounds travel fast and far in the ocean. We don't know exactly what they are saying, but we know what the calls might be communicating.

Whales use sound to:

- Find each other

- Find their prey

- Find their way around in dark waters

Other sounds are social and vary from whale species to whale species. Experts don't know what all their different calls mean, but they do know that bottlenose dolphins have special signature whistles they create shortly after they're born, which function as a kind of individual name for that dolphin. Other whistles might mean "Sharks are nearby!" or "Keep swimming and you'll run into a delicious school of fish!"

ECHOLOCATION

Narwhals, bottlenose dolphins, and other toothed whales produce high-pitched clicks to help them navigate and find food. It's called echolocation. They project a beam of sound through the bulbous part of their heads. It's a fatty structure, in front of the brain, that focuses on sound like a flashlight focuses light. When a clicking sound hits an object in the water, the sound bounces back to the whale as an echo. The whales use that to figure out the location, size, and shape of the object.

Scientists believe that the sounds bounce back and travel through the lower jaw to the inner ear and then are transmitted to the brain. That's how echolocation helps these sea creatures "see" using sound.

PREHISTORIC JELLYFISH

Jellyfish are like the ultimate hitchhikers. Whatever way the current flows, that's where our pal the jellyfish goes. For millions of years, jellyfish have been drifting through the world's oceans. They can be found in cold waters and warm waters, deep under the ocean's surface and in shallow waters along coastlines. Some are clear, but others are pink, yellow, blue, or purple.

Jellies are beautiful and scary. Their long, trailing tentacles can deliver painful stings.

Look, Ma, I'm Transforming!

For most of their lives, jellyfish look nothing like the bell-shaped creatures you see washed up on the shore. Jellyfish begin their lives as polyps. These baby jellyfish look like column-shaped creatures with a ring of tentacles around a central mouth.

Metamorphosis: a major change in the form or structure of some animals that happens as the animal becomes an adult

They attach themselves to a hard surface on the ocean floor and draw food into their mouths with their tentacles. As they grow, the polyps go through a process of metamorphosis. They begin by transforming into what looks like little stacks of tires.

When conditions are right, each little ring of tissue pops off and becomes a tiny jellyfish. This amazing process is called strobilation. This is really mind-blowing when you think about it. One single polyp transforms into many jellyfish! These tiny jellyfish quickly begin to feed and grow into the medusa form—the bell-shaped creatures with long tentacles we think of as jellyfish.

Did you know jellyfish aren't fish at all? They're invertebrates, meaning, unlike fish, they don't have a backbone. They also don't have a heart, gills, blood, or even a brain. Yet somehow they keep beating me at chess! Go figure.

Strobilation: a form of reproduction in which the body divides itself into segments, which later develop into separate individuals

SUPER COOL SCIENTIST

DR. REBECCA HELM

Dr. Rebecca Helm studies jellyfish at the University of North Carolina in Asheville so she can learn more about how jellies and other animals, such as frogs and butterflies, transform over time. Rebecca says jellies are really important to marine ecosystems, so understanding them can ultimately help us conserve and protect our environment.

Why Do Jellyfish Sting?

It's not because they love giving painful high fives. It's how they catch their prey. When a jellyfish stings a shrimp, the shrimp is stunned long enough for the jelly's tentacle to draw the shrimp to its mouth. It slowly moves the shrimp from its mouth to its stomach, where it is digested. Liquefied pieces of shrimp are then passed to different parts of the jellyfish. Anything that's left over, like hard bits of shell, goes back to the mouth to be spit out. So you could say that the jellyfish eats and poops through the same opening.

Note to self: Never let a jellyfish give you a kiss.

MOMENT OF

TO PEE OR NOT TO PEE

For years, people believed that the best way to take the burn out of jellyfish stings was to pee on them before scraping the stingers away. Unfortunately, this won't help with the pain, and it will just leave you with a stinging, pee-covered leg. Scraping away the tentacles doesn't work either. The pressure can make them release even more venom. One home remedy that does help is vinegar. Pour it over the sting and have someone (wearing gloves for safety) remove the stingers with a pair of tweezers. Next, apply heat to keep venom from spreading and doing more damage. Washing the area with salt water—not fresh water—will also help if there's no vinegar handy.

All jellyfish sting, but not all jellyfish sting people. Jellyfish have special cells in their tentacles that release venom, and that venom is targeted at the kinds of creatures they like to eat. We're not affected by the venom in jellyfish that eat shrimp, crabs, or other tiny creatures (because we are not shrimp, crabs, or other tiny creatures). You might be stung by one of those jellyfish and never feel it, because you don't have a lot in common with their prey. But when it comes to the types of jellyfish that eat fish—watch out! We often feel those stings, and they hurt!

To a jellyfish, humans and fish are pretty closely related. We're both vertebrates, meaning we have backbones. When a jellyfish accidentally brushes a human with its tentacle, the tentacle shoots out cells that look just like tiny needles from your doctor's office and injects venom into whatever's in the creature's path. This venom causes a hot, shooting pain that doesn't go away very easily.

SUPER COOL SCIENTIST

DR. GRACIELA UNGUEZ

Dr. Graciela Unguez is a professor in the Department of Biology at New Mexico State University, and she studies electric fish. Electric fish have thousands of tiny cells that produce electricity. These cells can be in their tails, their chests, or their stomachs. The most famous of these, the electric eel, generates up to 600 volts. That's sixty times more than a car battery! They use their jolts of electricity to immobilize—and then eat—smaller fish. Graciela wants to find out how electric fish can regrow all sorts of tissue in their bodies—even their spinal cords!

MEGA MATCHUP
OCTOPUSES VS. DOLPHINS

In this seafaring smackdown, we have two clever ocean creatures: one a cephalopod, one a mammal, both super cool. In one corner, octopuses! The eight-armed, three-hearted super shape-shifters. And in the other corner, dolphins! The speed-swimming, echolocating, super-jumping sea mammals. Which sea creature will come out on top?

TEAM OCTOPUS

- These incredible animals have been around for more than 296 million years—long before dolphins took to the sea.

- Octopuses have the most amazing camouflage on earth. They can change the color and texture of their skin to match the sandy ocean floor, pointy green coral, or spiky pink seaweed.

- With eight sucker-covered arms, giant eyes, a boneless body, and a beak for a mouth, octopuses are one of the most unusual and fascinating animals on the planet.

- Octopuses produce ink and store it in an ink sac. When a predator tries to strike—*whoosh!*—the ink cloud gives the octopus time to escape.

- Octopuses have nine brains—a main brain and a smaller brain for each arm—and three hearts—two to pump blood to its gills and another to send blood to the rest of its body.

TEAM DOLPHIN

- Fifty million years ago, hoofed land animals took to the sea and evolved into the smart, fun-loving dolphins we know and love. Their noses moved to the top of their heads and became blowholes, their front legs became flippers and their back legs grew into flukes—the dolphin's tail. How cool is that?

- A dolphin is more than just a pretty face. They're lean, mean swimming machines. Some species have been clocked at more than thirty-five miles per hour. They're epic jumpers too, leaping as high as thirty feet. That's as tall as a three-story house!

- Using echolocation, dolphins can tell the difference between a BB pellet and a kernel of corn from fifty feet away.

- Dolphins can heal themselves! Scientists don't understand how, but a dolphin that's been badly bitten by a shark can heal itself in a few weeks.

- A dolphin's brain-to-body ratio is the second highest in the animal kingdom; only humans have a bigger relative brain.

Which sea creature is cooler: Octopuses or Dolphins?

YOUR VERDICT

ANIMAL SUPERPOWERS

SUPER HEALERS

We all know what happens when you get a cut or scrape: You get a scab, you try not to pick at it, and then after a little while, you heal. Our human bodies have the ability to heal us up to a point. We can mend a paper cut, but we can't regrow an arm. That would be bananas. That would be…salamander territory.

These very special animals can regrow body parts! One specific type of aquatic salamander, the axolotl, is especially cool, not to mention supercute. These salamanders can regenerate missing limbs or tails, and even parts of their brains, hearts, and lower jaws.

Hey, Where's My Dinner?

If a heron grabs an axolotl by the tail, the axolotl can run away—"Adios, bird!"—leaving its tail behind. The heron is left wondering, "What happened?"

That's when the axolotl's body kicks into action. The wound is sealed just like in a human, but instead of a scar, something called a blastema forms.

After a few days, blastema cells turn into different kinds of cells—muscle cells, bone cells, nerve cells, blood cells. Those cells work together to begin to regrow the missing limb, just the way they did when the axolotl was growing inside an egg.

Blastema: a group of cells that have the ability to develop into a new body part

Scientists are studying axolotls to find out how humans can be more like these super salamanders. Imagine if we could grow back limbs, or livers, or hearts. With any luck, that's what our pal the axolotl can teach us!

REGENERATION NATION

- Starfish and newts regrow limbs.
- Sea cucumbers can regrow internal organs, like intestines, and heal deep cuts.
- Flatworms regrow their heads if they're cut in two.
- Deer regrow antlers.
- Lizards regrow tails.

What about growing extra limbs?

SUPER SLEEPERS

Circadian rhythms are like built-in alarm clocks. They keep our bodies on schedule. In humans, light from the sun gives us cues to wake up and go about our day. When darkness sets in, our brains know that it's time to feel tired. But what about those animals whose superpower is the ability to sleep for months at a time? (A power we are totally jealous of, by the way.) What kind of circadian rhythms are at work there?

PRONOUNCER
Circadian =
sir-KAY-dee-in

Circadian rhythms: the natural clock inside us that tells us when to do things like wake up, eat, and go to sleep

Arctic ground squirrels are one of the world's great hibernators. They hibernate in the winter, when there isn't enough food for them to eat. They're often asleep for seven to eight months out of the year. Bears also have a long hibernation period—five to seven months a year. But that's nothing compared with the edible dormouse, which can hibernate for eleven months.

A Whole Lotta Shaking Going On

No one needs to tell Arctic ground squirrels to chill out. They are experts in the subject.

Zzzzz

When they're hibernating, they can drop their body temperatures below freezing to match the outside temperatures of frozen winters. But even though their temperatures are way below freezing, they don't freeze into squirrel-cicles. That's because they use a process called supercooling. Every two to three weeks, without even waking up, the squirrels shiver and shake for up to fifteen hours to warm their bodies back to a normal temperature. When the shivering stops, their temperatures drop back down to below freezing.

Ha! I knew I smelled a squirrel on this page!

"IT'S TOO HOT. WAKE ME IN THE FALL!"

Some animals hibernate in summer. This is called aestivation, instead of hibernation. Desert tortoises, crocodiles, and some salamanders move underground, where it is cooler and more humid, to survive hot, dry summer months.

What Time Is It?

Scientists have learned that during hibernation, the circadian rhythms shut off and are replaced by the circannual cycle. It's a yearly rather than daily cycle. This cycle is what lets the animal know when to burrow underground and when to come out again. But exactly how bears and squirrels and other hibernators know when it's time to wake up and eat is still a scientific mystery.

SUPER SHAPE-SHIFTERS

If you were a spy with a super cool suit that allowed you to change both color and texture, all in less than the blink of an eye, your code name might be Agent Cuttlefish. Cuttlefish are masters of camouflage. These amazing sea creatures use a blend of three colors and two layers of skin to make themselves the ultimate shape-shifters.

Not a Fish!

The cuttlefish isn't a fish at all. It's a part of the cephalopod family, which includes

 MYSTERY PHOTO

Focus your eyes on this mystery photo. Can you guess what it is? Turn to the next page for the answer.

octopuses and squid. There are a lot of cool things about the cuttlefish, but its color-changing skin is the coolest. The cuttlefish's camouflage is among the most advanced in the animal kingdom.

Many people think that the plural of octopus is octopi, but it's actually octopuses. Octopi would be correct if the word were Latin in origin, like "cactus." But octopus comes from the Greek word *oktÐpous*, which means you make the plural by adding an "es" instead of switching to an "i." I love grammar! (Which comes from the Old French *gramaire*, and initially the ancient Greek *grammat-*, *gramma*, meaning "letter," but descending through ancient Greek, Latin, and then Old French as...)

Cuttlefish and other cephalopods have millions of tiny colored organs in their skin called chromatophores. They're like tiny sacks on the skin's surface, filled with color. They come in three shades: yellow, brown, and red. The cuttlefish can stretch or shrink its chromatophores, mixing colors to match its surroundings.

Chromatophore: a cell that contains pigment

But that's only part of it. The cuttlefish has a second layer of skin with proteins in it that can reflect light. It's like wearing tiny mirrors. Cuttlefish can manipulate those mirrors to reflect back the blues, greens, and even purples coming from the light underwater.

ANSWER!

It's a spider's foot! Most spiders have pads on their feet covered in tiny hairs. And these hairs split into even tinier hairs. It's all these tiny hairs that help spiders walk on walls. These subdivided hairs are shaped like spatulas at the end so they can have as wide a surface as possible touching and sticking to the wall. This stickiness is thanks to something called van der Waals forces. This force comes from the attraction between any two molecules. If you bring two molecules close enough, there's actually a little bit of a pull between them, like magnets. Normally you can't feel this attraction when you touch something, but these tiny hairs have so much surface area that there's enough force to let spiders walk on walls.

It's this combination of chromatophores, colors, and light-reflecting proteins that lets cuttlefish create almost any color. You might say they are camouflage artists.

As if that wasn't enough, cuttlefish take their camouflage to the next level thanks to papillae on the skin's surface. These little lumps of skin can change shape to become smooth or bumpy or spiky. That's some seriously smart skin.

And cuttlefish can change both their shape and color in just a third of a second! In the time it takes you to stick out your tongue, a cuttlefish can camouflage itself three times! How do they do so much, so fast? That's a question scientists are still trying to answer.

PRONOUNCER
Chromatophore =
crow-MAT-uh-fore

I've been in the laboratory-kitchen all day, and I finally did it! I invented Chromato-S'mores!

Where are they?

They're like regular s'mores—you know, chocolate, marshmallow, and graham cracker—but they change color so they blend in with whatever plate they're on. That way no one will know you're eating a plate of s'mores for dinner.

Can I try one?

Yes! As soon as I find it. I put it down somewhere, and, well...

SUPER STINKERS

Imagine going for a walk on a warm summer's night. The sun has just set, you're wearing your favorite sandals and shorts, and there's a perfect breeze blowing by. But wait a minute. What's that smell? Oh no. It couldn't be. But. It. Is. SKUNK!

Skunks can spray their predators with a foul-smelling chemical combination that doesn't just stink—it burns your eyes and chokes your lungs. Without this super defense system, slow-moving skunks wouldn't be able to escape from predators, or curious dogs.

But skunks aren't the only animals that use stink as a superpower to keep enemies away. Vultures eat the rotting flesh of dead animals. When a predator that is all too alive threatens a vulture, it lets the puke fly. The stink of vulture vomit is usually enough to scare off even the hungriest animals.

Opossums, on the other hand, freeze when they're threatened. They can lie perfectly still for hours, fooling foxes, coyotes, and owls into believing that they're dead and not edible. But

SUPER COOL SCIENTIST
DR. RICKY LARA

Dr. Ricky Lara at the University of California is an expert at what puts the stink in stinkbugs. That funky smell is a natural chemical that is part of the insects' defense system. Stinkbugs release their odor when predators are near. Surprisingly, Ricky tells us that stinkbugs taste a lot better than they smell. There are even recipes for stinkbug chocolate chip cookies and stinkbug salsa.

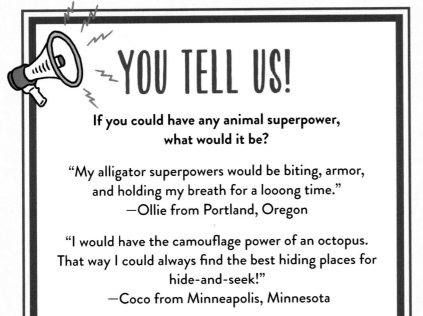

YOU TELL US!

If you could have any animal superpower, what would it be?

"My alligator superpowers would be biting, armor, and holding my breath for a looong time."
—Ollie from Portland, Oregon

"I would have the camouflage power of an octopus. That way I could always find the best hiding places for hide-and-seek!"
—Coco from Minneapolis, Minnesota

when one of those predators doesn't want to give up and go away, opossums get serious. They poop an extra-pungent green mucus—proof that sometimes the butt defense is the best defense.

SUPER PEEPERS

What do dragonflies call speed skating? Just skating. That's because dragonflies see faster than we do, making even our fastest skaters look as if they are moving in slow motion. Scientists have discovered that the smaller the animal and the faster its metabolism, the faster that animal sees. That's how insects like killer flies and dragonflies, which can fly as fast as thirty-five miles per hour, are able to hunt and swoop in on their prey without missing a beat.

Insects have photoreceptors in their eyes that allow them to see really quickly. These photoreceptors help them to be more aware of their surroundings and react in time to avoid a predator or catch their prey. So the next time you try to swat a fly and miss, don't feel so bad. You're competing with a little high-speed camera on wings. We're just slow-moving humans. But hey, we live way longer.

ANIMALS WITH HIGH-SPEED VISION

- Dragonflies
- Chipmunks
- Hummingbirds

Did you know we see the world around us as a nonstop video, but in reality we're piecing together lots of distinct images sent from our eyes to our brains? It's kind of like a flip-book. Humans see 60 images per second, turtles 15, and flies 250.

SUPER COOL SCIENTIST
DR. PALOMA GONZALEZ-BELLIDO

Dr. Paloma Gonzalez-Bellido at the University of Minnesota studies how insects see the world and how they catch their prey. She and her team raise killer flies and dragonflies in their lab to try to learn how these creatures hunt and kill while they're flying. How can she possibly study insect brains? She puts them under a microscope and uses tiny electrodes on the brains to measure electricity. One day, this research could lead to breakthroughs in improving human sight.

MEGA MATCHUP

PORCUPINE CARIBOU VS. MONARCH BUTTERFLIES

It's time for a battle between two long-distance travelers that migrate thousands of miles without a map or a compass for assistance. In one corner, porcupine caribou! The four-legged, fuzzy creatures that wander the Arctic in giant herds. And in the other corner, monarch butterflies! Royalty of the insect world, which fly thousands of miles to their breeding grounds. Which mega-migrator is the most magnificent?

TEAM PORCUPINE CARIBOU

- These super migrators can travel 3,000 miles every year across parts of Alaska and Canada.

- Caribou, also known as reindeer, are the only member of the deer family in which both males and females grow antlers. They use them to dig in snow for food. Awesome headgear for everyone!

- They have foot-tastic hooves that can spread wide to act like snowshoes in winter or like paddles when they swim.

- You think your car pool is crowded? These herds can be as large as 169,000 caribou!

- Caribou have special scent glands near their ankles. When there's danger nearby, they kick up those legs and release an odor that lets others know to watch out. Saved by smelly feet!

TEAM MONARCH BUTTERFLIES

- Tiny monarch butterflies can travel up to 3,000 miles to reach their winter homes in California and central Mexico.

- Monarchs are poisonous to predators! That's because as young caterpillars, they eat tons of milkweed, which is toxic. You are what you eat, after all.

- They can taste with their feet! It helps them know where to lay eggs. They simply land on a plant and taste it through their feet, and then they know if their baby caterpillars can eat it.

- Monarchs' wings are covered in tiny scales that overlap, kind of like shingles on a roof. These scales help the butterflies climb higher faster when they take off. They also lose these scales over time, so you can roughly estimate a monarch's age based on how many scales it has left.

- Monarchs make solo journeys down south, but once they arrive, they will cluster with other monarchs on trees. Some trees can be covered in tens of thousands of monarchs at once! That's a sight worth seeing.

Which super migrator is cooler: Porcupine Caribou or Monarch Butterflies?

YOUR VERDICT

MOMENTS OF UM

HOW DO FLYING SQUIRRELS FLY?

Flying squirrels don't really fly—they glide. They have flaps of skin between their legs, arms, and chest. When they leap off a branch with their arms and legs stretched out, it's like jumping with a parachute.

DO SEA MAMMALS DRINK SALT WATER?

All mammals need water to survive, but drinking salt water can make land mammals sick. How do sea mammals get the H_2O they need? It's a question that has puzzled scientists for years, but experts now believe that whales, dolphins, and other sea mammals need a lot less water than land animals to survive. They get most of that water from the sea creatures they eat, and their kidneys filter out the salt.

ARE ELEPHANTS REALLY AFRAID OF MICE?

Yes and no. Elephants don't have great eyesight, so it's not surprising that something that suddenly darts past them might be startling. But it doesn't have to be a mouse. Any small creature crazy enough to make sudden movements at the feet of the earth's biggest land animal might spook an elephant.

WHY DO SLOTHS MOVE SO SLOWLY?

Sloths move slowly for a couple of reasons. The first is for protection. If they're not moving, they can't be easily seen by predators like hawks and cougars. They're also conserving energy. Sloths' leafy diets don't give them a lot of nutrients like fat and protein, so their metabolisms are slow. They can slow down their metabolisms even more when they're swimming and can hold their breath for as long as forty minutes! Sometimes it pays to be slow.

HOW ARE CHEETAHS ABLE TO RUN SO FAST?

Cheetahs can run as fast as seventy miles per hour, making them the fastest mammal on land. Everything about their bodies is built for speed, from their strong legs and flexible spines to their enlarged lungs. They even have unique claws that help them grip the ground while they sprint—like built-in running shoes. Cheetahs' incredible speed makes it possible for them to chase down hares, impalas, and wildebeest calves during the day, while lions and leopards often have to wait until dark to hunt.

WHY ARE FLAMINGOS PINK?

Flamingos are actually born with white and gray feathers. They turn pink because their diets are rich in foods with pink pigment like shrimp, snails, and algae. Imagine if you were the color of your favorite meal!

WHY DO FROGS' TONGUES STRETCH SO FAR?

Frogs can snatch flies out of the air faster than a human can blink, thanks to their amazing tongues. Frogs' tongues are as long as one-third of their body length. If your tongue were that long, you could lick your belly button! Not that you'd want to—that sounds gross. Frogs' tongues are also ten times softer than a human tongue, which means they can stretch like putty.

HOW DO SNAILS GET THEIR SHELLS?

Snails are born with shells, but those shells look nothing like you would expect. Baby snail shells are very thin, and they look more like a glass slipper than a shell. Snails need calcium to make their shells hard and strong. A snail's first meal is usually the calcium-rich shell of the egg it hatched from. As the snail grows, its shell grows with it.

PART 2
PLANTS

FROM TINY SEED TO TOWERING TREE

HOW DO SEEDS GROW?

You think you know plants. They're green, they're leafy, they're boring, right? Don't be fooled. Plants are action packed! They send secret messages. They fight off foes. Some even eat meat! Take trees, for example—they survive against great odds to become the towering giants we love.

Growing into a tree is kind of like playing a video game—you complete one level and it's on to the next. So grab your controller and get ready to play:

Tree Quest
A Plant-tastic Video Game

Level 1:
Seeds Can't Run
You are a seed. You can't move. Your mission: don't get eaten or destroyed! This level is almost all luck.

Level 2:
Thrive to Survive
Time to power up by taking in water and eating your built-in supply of starch. Yum!

Level 3:
Roots and Shoots
Oh no! You're almost out of starch. Time to grow a root to soak up water and minerals from the soil. You'll also need to work on a stem, called a shoot.

Level 4:
Seed Starch is for Plant Noobs
You made a leaf! Now you can photosynthesize. That's when you take water, carbon dioxide, and sunlight and turn them into oxygen and sugar for you to eat.

Level 5:
Barking Up the Right Tree
Time to branch out aboveground and grow lots of roots belowground. Don't forget to customize your tree with fruit, bird's nests, and tire swings.

Unbe-LEAF-able:
YOU WON!
You did it! You're a tree! To play again, make a new seed and drop it on the ground. Good luck!

It's Raining Seeds!

There's nothing like sipping a cup of hot cocoa next to a cozy fire. It's the best. Unless you're a tree, because (1) trees don't drink cocoa, and (2) FIRES BURN TREES!!! But some fires are actually good for forests. You see, fires burn all the leaves and dead branches sitting on the forest floor. Once that stuff is gone, the ground is clear for new plants to grow. And some trees need fire to begin their growing cycle.

Giant sequoias have a thick bark—up to two feet deep—that acts like a kind of armor against fire. It's filled with a chemical called tannic acid that keeps the trunk from burning.

> Go, go, go! Find a patch of land.

Photosynthesis: the process by which plants and trees use sunlight to turn carbon dioxide and water into glucose and oxygen

So sequoias often survive fires just fine. But an interesting thing happens when the heat rises to the treetops. All sequoia trees have cones. Intense heat from the fire opens the cones, and millions

of giant sequoia seeds rain down from the canopy. Then wind scatters those seeds over great distances.

A single tree can have as many as 30,000 cones, with 200 seeds in each one. Only a small number will survive and level up to become a tree, though. Many land in a spot that's too dry. Others get eaten or trampled. But the few that do make it into the soil and ger-

Fact-astic Voyage

North American oak trees produce more nuts every year than all other trees in the region combined. But every two to five years, oaks produce way, way more acorns than they usually do. These are called "mast" years, and during these booms one oak can produce as many as 10,000 acorns. This huge surplus means that the chances are better for some of these nuts to turn into baby oak trees, rather than getting gobbled up by the animals that rely on acorns for their survival—like chipmunks, squirrels, blue jays, and even bears.

minate have hardly any competition, except from other sequoias, since everything else burned away.

Germinate: to begin to grow, or sprout

MYSTERY PHOTO

Focus your eyes on this mystery photo. Can you guess what it is? Turn to page 54 for the answer.

A Visit to the
Tree Hall of Fame

TALLEST
HYPERION

This towering redwood is 600 to 800 years old and stands 380 feet tall. That's higher than the Statue of Liberty! Scientists think it could have been even taller, but alas, some woodpeckers got to it first and damaged it near the top.

REDWOOD NATIONAL PARK, CALIFORNIA

OLDEST
METHUSELAH

This bristlecone pine is more than 4,800 years old—the oldest known tree on earth. It's older than the pyramids in Egypt! Its exact location is kept secret to protect it from tourists. This pine likes its privacy.

INYO NATIONAL FOREST, CALIFORNIA

LARGEST
GENERAL SHERMAN

This giant sequoia stands 275 feet tall and over 36 feet in diameter at the base, making it the largest tree by volume (which measures how much physical space it takes up). Its trunk weighs around 1,400 tons— the same as fifteen blue whales!

SEQUOIA NATIONAL PARK, CALIFORNIA

How Do Trees Spread Their Seeds?

Fruit trees and sequoias aren't the only ones with clever ways to spread their seeds. Here are some other plants that know how to travel in style:

- Maple trees drop seeds with helicopter-like wings so they can flutter away.

- Seeds from the cocklebur plant are covered in hooks that attach to fur to be carried to new ground.

- Dandelion seeds float away in the wind—like tiny hang gliders!

- Coconuts fall off trees and roll to a new patch of land. Some end up in the ocean, where they can float off to a tropical island to become a new tree. Nice way to start a life!

MOMENT OF EWW

POOP WITH A PURPOSE

There are many ways to travel. You can ride a bike or take a hike, fly in a plane or sit on a train. And it may sound absurd, but you can travel by turd. Or at least, some seeds do. Trees spend a lot of energy and nutrients making fruit, but it's worth it because animals love fruit, and fruit is full of seeds. The animals that love a tree's fruit help those seeds get around. Apples, for instance, are gobbled up by critters that will carry the seeds in their intestines to new locations. When they poop them out, those apple seeds will be surrounded by fresh fertilizer, so they can quickly start growing! Brilliant, right?

ANSWER!

It's a cocklebur! The burs—each of which contains two seeds—attach themselves with tiny hooks to passing animals like dogs, in an attempt to be dropped onto fertile ground and grow into new plants.

Historical Hoax

In 1894, a newspaper editor in Colorado came up with a fun idea to promote his friend's potato crop. They took a picture of a normal-sized potato, enlarged it, and pasted the new photo onto a board. Then they photographed farmer Joseph B. Swan holding the "mammoth" eighty-six-pound potato and printed the photo in the local newspaper. It wasn't long before *Scientific American* came across the photo, and it was printed in the magazine! By the time the editors realized it was a hoax, farmer Swan had been inundated with requests for seeds from the giant potato.

YOU TELL US!

Imagine you're a tree. How would you design a seed so it can spread far?

"I would create a seed that has a tiny little fairy inside, and once the seed was ready to fall from the tree, the tiny little fairy would come out, carry the seed, find a cozy little place to rest, and then *plop*."
—Izzie from Baltimore, Maryland

"I would use the classic propeller technique and rely on the wind to take the seed far from where it started."
—Julius from London, England

Anatomy of a Seed

Seeds can be as tiny as a grain of sand or as big as a coconut, but they all have three basic parts.

1. Embryo: a developing plant inside the seed

2. Endosperm: tissue that supplies the embryo with food

3. Seed coat: a covering that protects the embryo from being injured or drying out

Seeds need moisture and oxygen for the embryos to grow, push their way out of their seed coats, and begin to sprout.

Can Trees Talk to One Another?

Trees may look like the strong, silent type, but in reality they are constantly sending secret messages to one another. Warning about danger, preparing for attacks, talking smack about bushes for being so short. Okay, well, maybe not that last part. Trees do all this by sending chemical messages through the air and underground.

When certain kinds of trees are under attack by insects or other animals looking for a snack, they can send distress signals through the air to nearby trees in the form of scent molecules. Surrounding trees detect those molecules

Fact-astic Voyage

In 2012, a roughly 32,000-year-old plant was brought back to life by Russian scientists. The seeds of the flowering plant (scientific name: *Silene stenophylla*) were buried in Siberia by an Ice Age squirrel and found more than 30,000 years later under 124 feet of permafrost! The plant has delicate white flowers and doesn't look a day over 29,000 years old.

through their leaves and then flood those leaves with chemicals called tannins. Tannins make the leaves taste bitter and even make some attackers sick.

Some trees have also been found to connect underground using a network of fungi. Fungi need trees because they can't photosynthesize to make their own food (they're deep underground, after all, far from sunlight). So they often grow near tree roots, where they can take in some of the sugar that trees make naturally. In return, these helpful organisms make nutrients for trees, like phosphorus and nitrogen. That's teamwork! Plus, fungi can spread far and wide underground, like telephone wires

Fungi: Living organisms that are neither plants nor animals. Types of fungi include yeast, mushrooms, and mold.

in the soil. If one tree is in trouble, it can give off chemical signals that can travel from tree to tree through this fungal network. It's like a plant Internet, with more roots and fewer cat videos.

Roots Rule!

Over hundreds of millions of years, trees have evolved to get the moisture and nutrients they need from the soil and draw it upward to the branches and leaves in their treetops.

- Tree roots absorb water from the soil.

- Then, as moisture evaporates from a tree's leaves, there is a powerful suction effect inside that moves narrow columns of water up through the tree. It's kind of like water being sucked through a straw.

- Those straws carry water to every part of the tree. So the next time you see a tree, imagine it going *sluuurrrp!*

We may not be able to talk to the trees, but it's worth paying attention. Pick a tree and hang out with it. Notice what you smell. Look closely at the bark. Listen to the wind in the leaves. Trees may be quiet, but they're excellent company.

MYSTERY PHOTO

Focus your eyes on this mystery photo. Can you guess what it is? Turn to the next page for the answer.

ANSWER!

It's roots from a durian tree! Tree roots are covered with hundreds and hundreds of tiny hairs to help them absorb water. But trees don't just absorb water through their roots; they also release moisture back into the air through a process called transpiration. It's almost like exhaling!

Roots are also mighty! They're anchored in the ground to keep massive trunks from toppling over. They hold soil in place so it doesn't wash away in heavy rains. Over time, they can even break through sewer pipes and push up sidewalks. Roots are totally the biceps of the tree world.

LOOKIN' FINE, PINE!

Evergreen trees aren't actually ever—entirely—green. Coniferous trees like pines do lose their needles, but the cycle is just much longer than for trees that lose their leaves every fall. It can take two to three years before a needle falls off, but it will turn reddish brown before it does. And the needles don't fall all at once, so these trees always seem to be green.

PRONOUNCER
Coniferous =
kah-NIH-fuh-rus

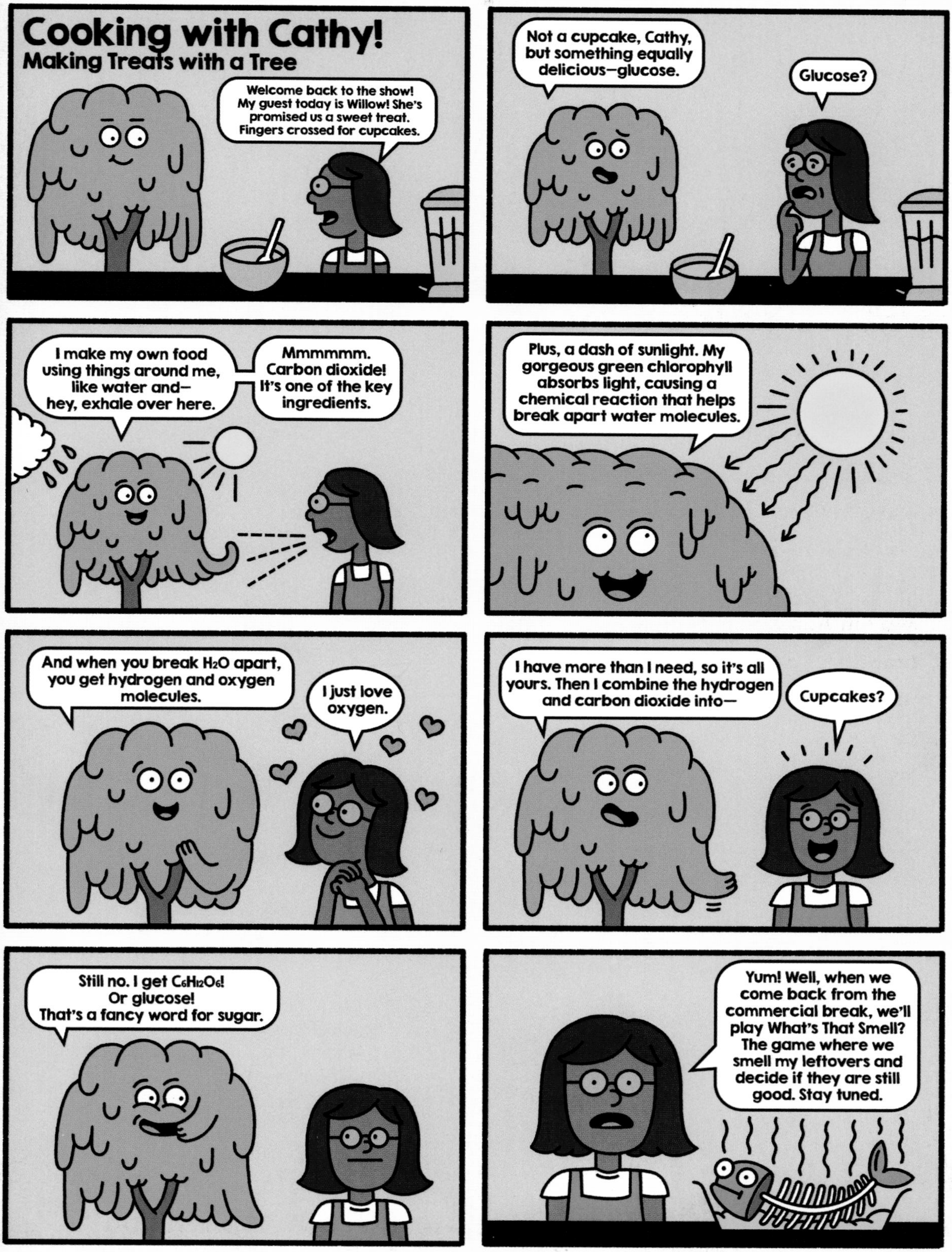
Cooking with Cathy!
Making Treats with a Tree
Welcome back to the show! My guest today is Willow! She's promised us a sweet treat. Fingers crossed for cupcakes.
Not a cupcake, Cathy, but something equally delicious—glucose.
Glucose?
I make my own food using things around me, like water and— hey, exhale over here.
Mmmmmm. Carbon dioxide! It's one of the key ingredients.
Plus, a dash of sunlight. My gorgeous green chlorophyll absorbs light, causing a chemical reaction that helps break apart water molecules.
And when you break H_2O apart, you get hydrogen and oxygen molecules.
I just love oxygen.
I have more than I need, so it's all yours. Then I combine the hydrogen and carbon dioxide into—
Cupcakes?
Still no. I get $C_6H_{12}O_6$! Or glucose! That's a fancy word for sugar.
Yum! Well, when we come back from the commercial break, we'll play What's That Smell? The game where we smell my leftovers and decide if they are still good. Stay tuned.

PLANTS: SURVIVAL OF THE FITTEST

HOW DO PLANTS FIGHT OFF PREDATORS?

Plants first showed up on land more than 400 million years ago with one goal in mind: TOTAL WORLD DOMINATION! Plants have spread all over because they are great at adapting—or changing—to survive. Like, did you know that plants developed their own circadian rhythms? That means their biological schedule keeps time with the cycles of the sun.

Some flowers open in the daytime and close at night. Sunflowers, for instance, not only turn toward the sun during the day; they turn east at night to be ready to catch the sun when it rises. It's a great way to soak up even more free sun energy!

Hi, new best friends. I'm Arabidopsis. I'm a plant!

I defend myself from insects by producing chemicals in my leaves that make bugs sick. But the really cool thing is, I know to make those chemicals during the day, when insects are most likely to attack. My defenses are set on a daylight timer! Take that, you bugs!

And don't feel bad about eating us. Veggies love being in salads. For us, it's like performance art.

My cousins cabbage, cauliflower, and broccoli do this too! But don't worry—the same chemicals that keep bugs away also help fight cancer in humans. So we protect ourselves from attack and make a healthy snack!

And guess what? We vegetables still follow circadian rhythms even AFTER we're picked from the ground. That means, even in your salad, I fill up with those chemicals in the middle of the day and have less at night. So gobble us up for lunch instead of a midnight snack.

For Crying Out Loud! Onions' Secret Weapon

There are a few things almost guaranteed to make you tear up: stubbing your pinkie toe, seeing the end of a Pixar movie, and chopping onions. The second you slice into an onion, it produces a chemical chain reaction that ends with you reaching for the tissues.

Chain reactions are super cool. One chemical reacts with another, and that creates a new thing, which then reacts with something else, and on and on—until eventually it creates something totally different from what you started with.

If someone tells you to eat your veggies to grow big and strong, believe them! Titanosaurs were the largest land animals *ever,* and these gigantic, long-necked dinos were total vegetarians.

SUPER COOL SCIENTIST

DR. JANET BRAAM

Dr. Janet Braam is a professor of biosciences at Rice University in Houston, Texas, where she studies the unexpected and surprising abilities of plants. She told her son about how some plants boost cancer-fighting chemicals in the daytime to defend against bugs. He replied, "Well, now I know what time I should eat my vegetables!" This inspired her to see if veggies still do this after they've been picked from the ground. Turns out they do! It goes to show, even a small comment can be the seed of a big idea.

MOMENT OF

WORMS: SUPERHEROES OF SOIL

When worms burrow through the soil, they're making space for air and water. It's great for plants, but what worms leave behind is even better—poop! Known as *castings,* worm poop is a superfood for plants. Worms eat organic matter like dead leaves, so their poop is filled with microbes that can help plants grow and fight diseases. It also improves the soil's ability to absorb moisture. Worm poop can even remove heavy metals and other toxins from the soil. Imagine so much magic from such a tiny butt!

When you cut into an onion, you're causing a chain reaction.

Chain reactions are so beautiful. They always make me tear up. In this case, literally.

The knife kicks it off by cutting open the onion's cell walls. Inside those cells are special enzymes. Once they're free, they interact with the air to create an acid. Another enzyme from the onion rearranges the acid to create *syn*-propanethial *S*-oxide. This tiny compound with the very long name is the eye-irritating, tear-inducing troublemaker here.

Enzyme: a substance made by living things that helps kick off chemical reactions

Onions are alliums, just like garlic and leeks. The chemicals in these plants act as a natural defense that keeps them from being eaten. Unlike us, animals don't have to cut into the skin of an onion to know to stay away.

HOW TO KEEP THE TEARS AT BAY WHEN YOU'RE SLICING ONIONS

- Make sure your knife is sharp. Dull knives open more cells, and that means there will be more chemicals that burn your eyes.

- Run the onions under cold water or stick them in the freezer for about fifteen minutes before chopping. Cold helps slow down the chemical reaction.

- Or you can wear goggles!

KITCHEN MYTHS: TECHNIQUES THAT
DON'T
KEEP THE ONION TEARS AWAY

- Chewing gum or bread
- Holding a spoon in your mouth
- Putting lemon juice on your knife

SUPER COOL SCIENTIST

DR. RABI MUSAH

Dr. Rabi Musah teaches chemistry at the State University of New York at Albany and studies the ways in which plants, like onions, use chemicals to defend themselves. She hopes that her work can help protect crops in ways that are healthy for the environment. Many plants produce molecules that make them unattractive to herbivores. If we know what these molecules are, we can use them to protect the plants on farms, instead of using unnatural chemicals that could hurt the ecosystem.

SURVIVING THE DESERT AT JOSHUA TREE

Picture a desert. Are you seeing sandy dunes? Dry, dusty rocks? Or chocolate cake? If it's that last one, you are picturing dessert, not a desert. Try again. Even though deserts seem too hot and dry for life, plenty of plants and animals live there.

To find out how they do it, let's take a journey to Joshua Tree National Park in California, one of the fiercest environments on the planet.

Joshua Tree National Park can go months without rain, and it can get as hot as 120 degrees Fahrenheit in the summer. In the winter, temperatures can drop below freezing. The plants

MYSTERY PHOTO

Focus your eyes on this mystery photo. Can you guess what it is? Turn to the next page for the answer.

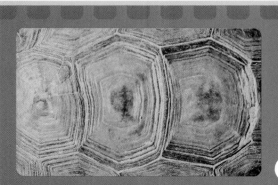

ANSWER!

It's the shell of a desert tortoise! These herbivores get water by munching on water-rich plants. They can store that liquid in their bladder for up to a year! Imagine holding your pee that long. Yikes! They also dig holes in the ground to sleep away the cold winter months.

Oasis: an area in a desert where water and plants are found

that live there have developed special skills to survive this extreme weather.

Some plants grow near a hidden pool of water, called an oasis. One way an oasis forms is when water from far underground seeps to the surface through cracks. The palms that grow there are perfect homes for small desert animals, like bats, mice, and birds. Larger animals, like mountain lions, coyotes, and owls, come by to get a drink and snack on the smaller animals. An oasis is sort of like a convenience store of the desert.

Some desert plants have waxy coatings on their leaves to help keep moisture in. Others open flowers only at night so they don't dry out in the sun. Plants like cacti collect as much water as they can when it rains, then store it in their bodies for the long, dry months. Their sharp spikes keep away animals that would like to get at that water. Great for defense, bad for hugs.

GRUESOME GARDENS

Lots of plants get eaten by bugs. But sometimes the tables are turned. Carnivorous, or meat-eating, plants evolved in places where the soil doesn't have the nutrients the plants require. So they developed a way to chow down on their natural enemy: insects! Get ready to enter the...

I'm not falling for that...again.

FREE HUGS

PLANT HOUSE OF HORRORS!

VENUS FLYTRAP

Brush just two of the tiny hairs on this plant's leaves, and it will snap shut like a clamshell! That's how it traps flies and spiders. Then digestive juices, similar to our stomach acid, melt down the bug so the plant can absorb its nutrients.

FOUND IN THE WETLANDS OF NORTH AND SOUTH CAROLINA.

SUNDEW

These alien-looking plants have leaves that are covered in sticky, pin-like tentacles. If an ant touches one, the tentacle will snatch it up at a speed four times faster than the blink of an eye and bring it to the plant's center. There, a glue-like substance holds the ant so the plant can feed.

MOSTLY FOUND IN AUSTRALIA AND SOUTH AFRICA.

PITCHER PLANTS

This plant looks like a bunch of colorful jugs...of DEATH! They lure in unsuspecting insects with sweet nectar. Once inside, the bugs slip down and down into a pool of digestive juices. Pitcher plants aren't picky eaters: Some species have been known to trap lizards, frogs, and mice!

FOUND IN BOGS FROM NORTH AMERICA TO CHINA TO AUSTRALIA.

SUPER COOL SCIENTIST
RICKY GARZA

Ricky Garza is a horticulturist and gardener at the Minnesota Landscape Arboretum, where he grows all kinds of weird, wonderful plants like the Venus flytrap and the super-stinky voodoo lily. Ricky also works to restore and preserve natural areas to make sure these interesting plants still have homes in the wild.

PRONOUNCER
Carnivorous =
kar-NIV-er-us

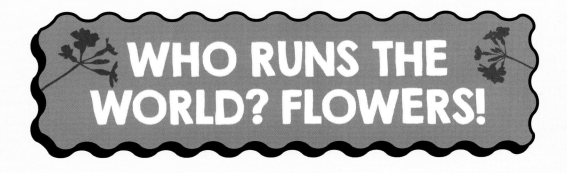

WHO RUNS THE WORLD? FLOWERS!

WHERE DID FLOWERS COME FROM?

For most of earth's long history, there were exactly zero flowers. Ice ages came and went, mass extinctions happened, dinosaurs started roaming the planet—but still, no flowers. If a stegosaurus wanted to impress another stegosaurus, it could give a bouquet of…ferns, maybe? Finally, somewhere around 140 million years ago (or maybe even longer), flowers burst onto the scene! They spread quickly and developed into all sorts of shapes and sizes. Everyone loved them (except maybe the other plants, which were probably jealous).

Today, flowering plants, also called angiosperms, make up at least 80 percent of ALL PLANT LIFE ON EARTH! They also give us fruits, grains, and vegetables. But where did they come from? And how were they able to spread so fast? Scientists still aren't sure. One flower that may have answers is *Amborella trichopoda* (pictured below). It's a shrub

with tiny white flowers that grows on an island called New Caledonia in the Pacific Ocean. This line of shrub has been around longer than any other flowering plant on the planet. By studying it, we might unlock the sweet-smelling secrets of all flowers' mysterious past.

Angiosperm: A type of plant that has flowers and makes seeds enclosed in a fruit. This group includes shrubs, grasses, and most trees.

Passing the Pollen

Itchy eyes, runny nose, super sneezes…are you sick, or is it just pollen? Pollen is a pain for us humans (especially in the springtime), but it is absolutely essential for flowering plants. Without it, flowers would be kaput.

You see, flowering plants can't make fruit or seeds until they are pollinated. That means pollen—this microscopic material full of genetic information—needs to get passed around. It's sort of like a very small, very important package that helps

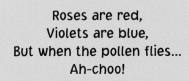

Fact-astic Voyage

Vegetables are angiosperms. That means plants like broccoli and brussels sprouts actually make flowers and produce fruit. We just don't eat their fruits. Instead, we eat other parts of the plant. With brussels sprouts, we eat the plant's buds, and with broccoli, we eat the flower itself before it blooms.

plants make seeds. Some flowers can self-pollinate, meaning the pollen doesn't have to travel very far, just from one flower part to another or from a different flower on the same plant. But other flowers can be pollinated only by pollen that comes from an entirely different plant of the same species. Unfortunately, there is no pollen Postal Service, so these plants have to get creative.

Some flowers pass their pollen by simply tossing it into the wind and hoping for the best. Sometimes it lands in a flower of the same species, but a lot of the time it doesn't. Instead, it might land in your nose and give you allergies! (For more on allergies, go to page 88.)

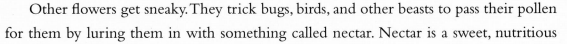

Roses are red,
Violets are blue,
But when the pollen flies…
Ah-choo!

Other flowers get sneaky. They trick bugs, birds, and other beasts to pass their pollen for them by luring them in with something called nectar. Nectar is a sweet, nutritious liquid found in flowers. The flower's bright petals and alluring smells are like billboards designed to catch the eye of passing animals. Those animals then stop and sip on this all-natural energy drink—and often, they accidentally get pollen on their feet or face. When they swing by another flower for more nectar, that pollen will come off and pollinate the second flower. Bada bing! That flower is now ready to start growing fruit and seeds!

CAN FLOWERS HEAR?

Flowers don't have ears, but they might be listening to sounds around them. That's what scientists learned from studying a type of flower called an evening primrose. When they played it bee buzzing sounds or similarly low-pitched buzzing sounds, the plant would quickly make more nectar. But when they played it higher-pitched sounds, nothing changed. The researchers think the flowers might be able to detect a bee's buzz and pump out more nectar to lure it in. When asked how they did this, the plants said...well, nothing. They might be able to hear, but they still can't speak!

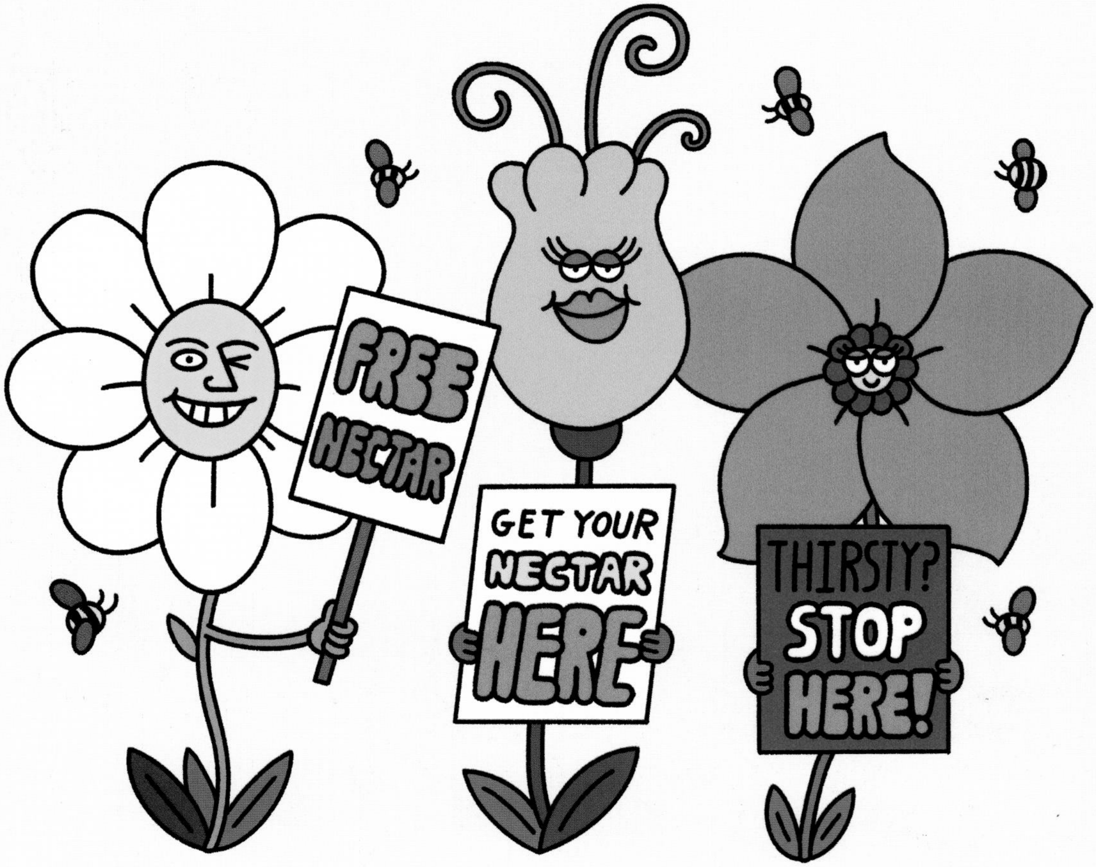
FREE NECTAR
GET YOUR NECTAR HERE
THIRSTY? STOP HERE!

Nature's Tricksters

Getting a bug to pick up your pollen is super important for many flowers. In fact, some flowers can be pollinated only by a very specific fly or wasp. Others can be pollinated by a wide range of bugs—but they need those bugs to find them first. So many flowers have developed clever tricks to catch the attention of an animal.

- The Solomon's lily flower gives off the scent of rotten fruit, a smell that vinegar flies go bananas for!

- Bucket orchids attract bees with sweet smells. Bees then slip into their bucket-shaped lips and have to crawl out, getting covered in pollen along the way.

- Several types of orchids smell like—and even look like—female wasps. Male wasps come searching for love but end up covered in pollen.

- The parachute flower smells like a dying bee, luring in certain flies that like to eat the juices made by dying bees. Gross but clever.

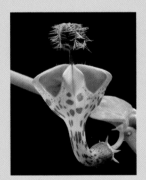

THE ROSE THAT WENT TO SPACE

Even astronauts like roses. In fact, NASA brought a miniature rose plant on a space mission as part of an experiment. They wanted to see if it would grow differently far above Earth. Turns out, it did! The lack of gravity changed its smell and made it lighter and crisper, according to NASA. It was so striking, perfume makers started using its scent in a fragrance. No wonder that variety of rose is called "Overnight Scentsation."

MEGA MATCHUP

DURIAN VS.
CORPSE FLOWER

Are you ready to take a great big whiff of two malodorous yet magnificent plants? In one corner, we have a fruit so smelly it's banned from public transport in many Asian countries—the durian. And in the other corner, a flower that smells like a dead animal—the corpse flower. Which one will come out on top as the greatest stinker?

TEAM DURIAN

- Durian fruit is called "the king of fruits." It's about the size of a cantaloupe and has a hard, spike-covered shell with a soft, custardy inside. It tastes delicious and can be used in all sorts of recipes, like in curries, candies, and cakes (to name just a few).

- Besides being delicious, it has a very strong and distinctive odor that is sometimes compared to smelly socks.

- Durians developed their powerful scent to attract animals like orangutans, elephants, and rhinos. These animals gobble up the fruit and poop out the seeds, creating even more durian trees.

- Flying foxes, also called fruit bats, are huge creatures with a wingspan of up to five feet. They LOVE durians. They're the main pollinator of the fruit.

- Recent studies have shown that one of the durian's closest ancestors is the cacao—the plant that gives us chocolate!

TEAM CORPSE FLOWER

- The corpse flower is famous for its odor—a combination of stinky cheese, garlic, rotting fish, and smelly feet. In other words—it smells like a dead animal.

- The smelly plant is also famous for its enormous flower spike, called a spadix, which can grow as tall as twelve feet.

- When the corpse flower unfurls, it emits a powerful, rotting-meat stench. The smell, along with the color of the flower, attracts pollinators like flies and carrion beetles that feed on dead animals. The plant heats up to help spread the odor particles around.

- It takes seven to ten years for these natives of the Sumatran rain forest to gather enough energy from the sun to bloom for the first time. After that, they flower every four or five years. Once open, the flower blooms for only twenty-four to thirty-six hours before it collapses and disappears.

- Corpse flowers are basically rock stars when they bloom. Because the event is so rare, people line up at botanical gardens to smell it in action.

Which stinker is cooler: Durian or Corpse Flower?

YOUR VERDICT

MOMENTS OF UM

WHY DO BANANAS MAKE OTHER FRUIT RIPEN FASTER?

Bananas—and many other fruits, like apples, pears, and mangoes—produce a gas called ethylene. This gas stimulates the ripening process in the banana itself and in other fruits nearby. When this happens, chlorophyll (the green pigment in plants) starts changing into other compounds—which have red, orange, or yellow colors. Starch gets broken down into simple sugars, so fruit becomes sweeter.

IF YOU PLANTED A POPCORN KERNEL, WOULD IT GROW A POPCORN PLANT?

If you take a kernel of popcorn and put it into the ground, it will grow into a corn plant. If you want to make popcorn, you'll need to take a corn cob off the plant, remove the kernels, and dry them out. Then you can take those kernels and make popcorn! You can't do this with the kind of corn you eat right off the cob, because only popcorn kernels will pop up into a delicious snack.

HOW DO PLANTS GROW UNDERWATER?

Plants are smart! Marine plants have evolved to absorb the moisture and carbon dioxide from the water they live in. They grow in the top levels of oceans, lakes, and rivers to get the sunlight they need to survive.

WHY DO FLOWERS SMELL GOOD, AND WHY DO THEY COME IN SO MANY COLORS?

Flowers smell good for the same reason the durian fruit and the corpse flower have such potent odors: to attract insects and birds that will spread their pollen and fertilize their flowers. Bright-colored flowers have a similar effect—acting like colorful signs saying, "Hey, bugs and birds, come check me out!"

WHY ARE DANDELIONS CONSIDERED A WEED?

Dandelions were once prized as one of the most beautiful flowers on the planet! The flowers and leaves are more nutritious than a lot of vegetables, and tonics made from the plant were used to fight a whole range of ailments from upset stomachs to skin rashes. And who doesn't love blowing on a dandelion puffball? In the twentieth century, humans decided they liked their green lawns more than dandelions. It wasn't long before the dandelion, which can grow just about anywhere, was considered a nuisance that needed to be destroyed.

WHY DOES MOSS GROW ON TREES?

Moss is super cool. It can grow anywhere that's moist and shady, like on trees, rocks, and fallen logs, and it was one of the first plants to grow on earth. Instead of roots, mosses have something called rhizoids that hold the moss in place and often collect water and nutrients. The plant is so good at absorbing moisture that battlefield doctors in World War I, who were worried about running out of bandages, used moss to treat wounds instead.

WHY DO RASPBERRIES HAVE TINY HAIRS?

Because they forgot to shave! Just kidding. When you roll a raspberry between your fingers, it'll come apart into tiny red jewels. That's because each tiny red jewel is its own little fruitlet, which means each one has its own seed and its own flower parts. Those little hairs are pistils, the female part of the flower. When the flower was in bloom, it attracted bees to transfer pollen, and the seeds developed into fruit.

DO PALM TREES HAVE RINGS?

No, palm trees don't have rings in the inside of their trunks—which means it's hard to tell how old a palm tree is. This is because they don't produce cambium like other trees, and their trunks don't grow as wide as those kinds of trees do. It turns out that palm trees aren't in the same plant group as most leafy trees. In fact, palm trees are more closely related to grass and bamboo than to leafy trees or conifers!

PART 3
HUMANS

GREETINGS FROM BODYLAND!

Did you know that the world's finest theme park is *inside* you? It's full of spills, chills, and spectacular thrills. What is this place, you ask? Bodyland! The only amusement park where it's totally normal for everything to be covered in blood!

THE NUTS AND BOLTS OF YOUR BLOOD AND GUTS

Since you can't see what's inside a body, you've probably never given someone a compliment like, "Hey, nice spleen," or "What a lovely gallbladder," or "Oh my, where *did* you get that pretty pancreas?" But all that stuff is hugely important and deserves to be celebrated! Your body is made up of several systems, each with its own rides, games, and important functions. You can visit each one and find out how they all work together. But to get the true Bodyland experience, you would have to spend a solid week here. So consider this just the tip of the system iceberg. Let's start with the heart of the whole operation…

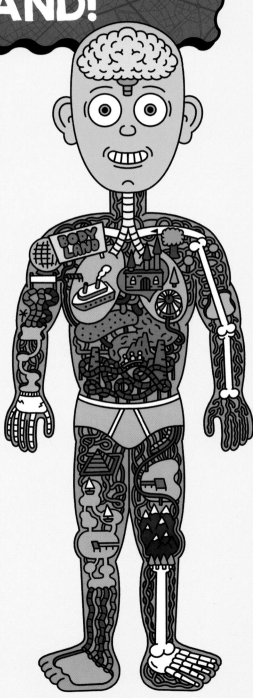

Ready to Get Pumped?

Welcome to the circulatory system! Hold on to your hats, because it's a wild ride. From here, you can travel to just about anywhere in the body. At the center of the circulatory system is…the heart! It uses a series of small tubes, called arteries, to move blood from the heart to the rest of the body. Blood is super important because it carries oxygen, something every part of your body needs. So your circulatory system helps move it all throughout Bodyland. When that blood is low on oxygen, veins move it back to the heart to start the cycle again.

The heart is a muscle made up of four separate areas, called chambers. The two chambers on the right take in blood that has just come back from circulating around the body and send it to the lungs, which give it some fresh oxygen. Then the two chambers on the left take that freshly oxygenated blood and send it back out into the body.

You know that *lub-DUB* sound your heart makes when it beats? *Lub-DUB, lub-DUB, lub-DUB.* Those sounds are made by the valves opening and closing between the different chambers in your heart. It's the

My heart is an excellent drummer. Too bad my spleen stinks at guitar.

valves' job to make sure blood is going where it's supposed to go. The first sound is the *lub*—it's quieter and made by the valves that sit between the bottom and top chambers on each side of the heart. The second sound—the *dub*—is coming from the valves that open to the rest of the body. They have higher pressure because the blood has farther to go—and so that sound is a little louder. *Lub-DUB, lub-DUB, lub-DUB.*

FOUR CHAMBERS OF THE HEART

THAT THUMP-THUMP IS A BLOOD PUMP.

1. **Right Atrium:** This chamber takes in blood that's low in oxygen, also called deoxygenated blood, then moves it to the...

2. **Right Ventricle:** Blood moves from this chamber into the lungs, where it can pick up loads of oxygen.

3. **Left Atrium:** Now, full of oxygen, that blood moves back from the lungs into this chamber.

4. **Left Ventricle:** This chamber pushes blood back into the body for another trip around Bodyland. Remember, keep your hands and feet inside the ride at all times!

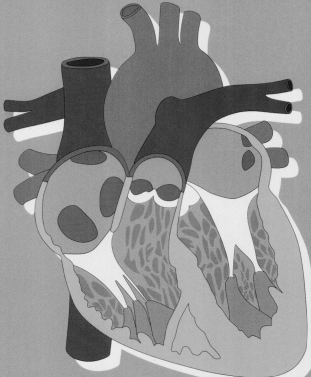

MOMENT OF
EWW

SLITHERING SNAKE SURGERY!

Looking at a snake's sliced-open beating heart might sound like a scene from a horror movie, but that's the way a scientist named William Harvey found out how hearts circulate blood through our bodies. He lived about 400 years ago, and until then, most people thought lungs moved the blood around the body. This was centuries before we had machines that could see inside us, so William did the only thing he could to satisfy his curiosity. He carefully cut open a living snake so that its heart was exposed. And he noticed that when he pinched off the vein below its heart, the heart started to shrink and grow pale. This showed that the veins are responsible for bringing blood back to the heart. The opposite was true when he pinched an artery. The heart began to fill with blood, because William had shut off the escape route.

O You Are 2 Much!

As you can see, Bodyland is an exciting, exhilarating, extra-fantastic place. If you need to take a moment and let the fact that you are actually here sink in, go ahead. Take a deep breath. Hold it. Now, exhale. Ahhhh, relaxing, right? And since you're already breathing, it's the perfect time to head over to the next system in the park: the respiratory system.

The respiratory system is responsible for bringing in oxygen (also called O_2) and getting rid of a gas called carbon dioxide. *Respiratory* comes from the word "respiration," which is a fancy way to say breathing! When you breathe, you pull in air through your mouth or nose, down the windpipe in your throat, and into the respiratory system's star performer: the lungs. That's where oxygen gets absorbed into the blood and pumped by the heart out to your body (thanks, heart!).

All your cells need that oxygen—plus something called glucose—to make energy. That energy powers you and your Bodyland, from head to toe.

Glucose: a type of sugar that many living things, including humans, use for energy

When cells in our body are done chowing down on glucose, they leave something behind: carbon dioxide. This gas is put back in the blood and circulated to the lungs for us to breathe out. Phew! See ya later, carbon dioxide!

THE HIGHS AND LOWS OF AIR PRESSURE

Don't freak out—but right now, there is air pushing on you from every single direction. It's got you surrounded! You don't feel it because it's always there. We call this air pressure. And it's actually a good thing, because it means there's always air to breathe. But did you know the air pressure around you changes depending on your altitude—that is, it changes depending on how high up you are? The higher you go, the less air pressure there is. This is important, because less pressure means fewer oxygen molecules for you to breathe in. So when you're high up on a mountain, there's less air for yodeling, blowing up balloons, and, most important, breathing.

Less oxygen up high? Is that why we say the view from a mountain is breathtaking? (cough)

All this breathing, blood pumping, and cell feeding happens without us ever thinking about it. But we can consciously control our lungs too. Like when you take a deep breath, then blow out some birthday candles, you're filling the lungs with more air than usual and making them push it out with much more force. Did you make a wish?

Fact-astic Voyage

There are hundreds of millions of tiny, spongelike cells called alveoli that line the inside of the lungs. They are responsible for getting oxygen into the blood. If you could flatten and lay out all the alveoli in just one lung, they would cover a tennis court.

I wished for more birthdays. One a year is not enough.

The diaphragm is located toward the bottom of the ribs, just below the lungs. It's the muscle that helps pull and push air in and out of the lungs. When we inhale, it goes down, and when we exhale, it pushes up. It's mostly on autopilot, making you breathe. But you can control it too, like when you blow out birthday candles or sing!

PRONOUNCER
Diaphragm =
DYE-uh-fram

Do You Hear That Rumbling?

Sounds a bit like growling. Maybe some gurgling and, uh-oh, was that a fart? Hmmm… these noises must mean you have reached Bodyland's digestive system. Yep! It's the system that takes you on a journey to break down food, provide nutrients to the body, and get rid of waste. Don't forget to put on a complimentary Bodyland water poncho before you hop into the inner tube. This system is going to have you sloshing around in the digestive rapids, and it's gonna get messy.

Your ticket for this ride? Just a piece of food. Pop it into your mouth and let's go! Your mouth starts breaking it down with those big, beautiful chompers known as teeth. *Nom nom nom!* Watch out! Here comes the saliva! *Splurt!* It's full of enzymes that help break down that food.

Hold on tight, because there's a giant drop next. One big swallow and you're in the esophagus—*whoosh!* At the top of the esophagus is a sphincter—it's a circular muscle that opens and closes to let different particles pass through. And the esophagus is lined with muscles that push food down and keep it from coming back up. At the bottom of the esophagus is another

sphincter—this one is the gate to the stomach. Make sure your poncho hoods are up!

I love it when words use the letters "ph" to make an "f" sound. It's PHully my PHavorite!

Gastric Fantastic

As food makes its way through the stomach, it is dunked, sprayed, and massaged with a combination of acids called gastric juice. The acid is so potent that if you were to touch it, your skin would burn. Good thing your stomach is lined with a special, snot-like mucus that protects it from gastric juice. And good thing we can't see inside our stomachs, because all this sounds disgusting.

Splash! Now the piece of food is getting sprayed with acids and squeezed by the muscles in the walls of your stomach. It's pretty gross. Gross enough to make you want to barf, which makes sense because this is where barf comes from! Yep, this half-digested stomach goop is the same stuff that comes up when you hurl. Muscles in the diaphragm and abdomen help out a bunch too. To launch your lunch, these muscles contract, squeezing your stomach and forcing its contents up and out. Speaking of which, if you are feeling queasy, there is a barf bag under your seat.

GET TO KNOW YOUR SPHINCTERS!

Sphincters are located all over the body. There are not one, not two, but SIX sphincters in the digestive system. You can think of them as gatekeepers between different digestive zones. These important muscles do all sorts of things like keeping food from going where it isn't supposed to and making sure stomach acids don't burble up into your esophagus. And, of course, don't forget the sphincter at the end of this great trip through the digestive system: the one that opens and closes to expel ~~stool feces waste~~...poop!

The Small, the Large, and the Stinky

Once your food's done breaking down in the stomach, it's ready for the next phase of this trip: the intestine safari! Hopefully, you brought a flashlight, because that meal you ate is now winding through a series of long, dark tunnels. Oh, and keep your eyes peeled for wildlife! The intestines are filled with strange and wonderful creatures known as gut microbes. These tiny bacteria help you digest food and are the reason you fart.

The journey starts in the small intestine. At this point, those chicken fingers or that apple you ate is a big, sloppy mess of goo called chyme (rhymes with time). The small intestine absorbs all the important vitamins and nutrients from that chyme and whisks them off to different parts of the body. This is how you get the nutrients you need to do everything from keeping your bones strong to feeding your cells with glucose. In fact, the small intestine is where 90 percent of nutrient absorption happens.

Next, we enter the large intestine. It's a tube that snakes up and around the outside of the small intestine. Its job is to soak up more water from the chyme, turning it into a drier, thicker lump. Bacteria help here too—in fact, some bacteria eat stuff in your chyme and release vitamins, like vitamin K, for your body. Isn't that nice of them? Finally, your food is ready for the rectum, which means the safari is nearing the end. And by the end, we totally mean the butt.

The rectum is a waiting room of sorts. Stuff sits here until it's ready to exit your body. At this point, the meal from several hours ago looks nothing like how it started. It's transformed, like how a caterpillar transforms into a butterfly. But in this case, it transformed into poop. Some of that waste is stuff your body just can't break down, some of it is water, but A WHOLE LOT of it is actually bacteria. In fact, 50 to 80 percent of the solids in a typical turd is made of bacteria! And with that funky fact, we've reached the end of the ride. Please exit through the anus. Bon voyage!

ALL SYSTEMS SECURE!

What happens when a gang of pesky parasites or vile viruses tries to sneak into Bodyland? They get stopped at the gate, of course. Bodyland's gate is your skin. It's the first line of defense against bad bacteria and all that other harmful stuff. Plus, it looks cool—kind of like body wrapping paper!

Have you ever scratched an itch and seen some flakes fly off your skin? Those are the dead skin cells that make up the top layer of your skin. You're covered in them. But

MOMENT OF EWW

A KERNEL OF TRUTH

You know how after you eat corn, you can see whole kernels embedded in your poo? Well, this corn-founding phenomenon is thanks to the outer shell of the corn kernel, called the hull. It's made of cellulose, a rubbery substance found in plants. Our bodies don't produce an enzyme that can break down cellulose, so it looks the same even after passing through the digestive system. But looks may be deceiving. Your body can break down all the corn stuff inside the hull, so that little yellow thing that came out of your backside? That's a hull filled with poo. But please, just trust us. No need to check for yourself.

there's so much happening with your skin that you can't see. Under that layer of dead cells, there's a whole lot of work being done to keep you healthy and safe.

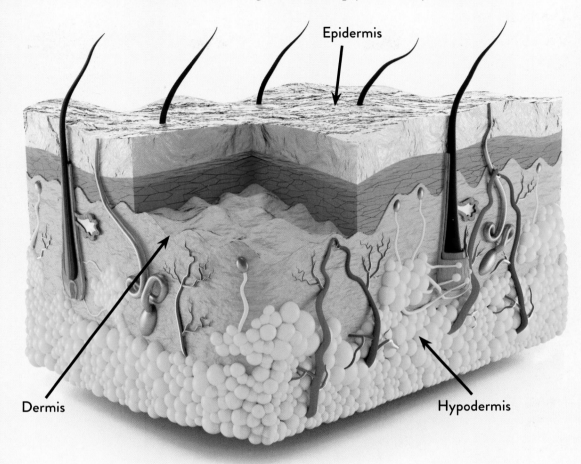

- **Epidermis:** This top layer of skin makes melanin—the pigment that gives color to your skin—and new skin cells. Don't forget that the epidermis makes you waterproof too!

- **Dermis:** This is the second layer of skin, where your nerve endings are—those help you feel things like a pinprick or the soft fur of a cat. Also in the dermis are the follicles that grow hair and all the blood vessels that feed your skin.

- **Hypodermis:** This third and final layer is made up of fat and other tissues. Fat is key to keeping your body warm on cold days. Can you guess which parts of the body have the thickest hypodermis? That would be the soles of the feet, the palms, and the butt.

Feeling overheated from your workout? Don't sweat it! Actually, do sweat it! During exercise, muscles burn up energy and give off heat. Sweat is the body's response to help get rid of this heat. The dermis is covered in sweat glands—little tubelike openings that fill up with sweat. Once full, the sweat seeps out and evaporates into the air, taking some of that heat with it and cooling you off.

Fact-tastic Voyage

Your skin is the largest organ in your body. It's made up of about 35 billion skin cells, and they are constantly being refreshed. You get a new set of skin cells every twenty-seven days. An adult's skin, weighed on its own, would be somewhere between eight and twenty pounds.

EVEN THE GATES ARE SYSTEMS

That's right, you thought you were just looking at the gate of Bodyland, but skin is actually part of another system—the integumentary system. In addition to skin, this system is made up of hair, nails, and exocrine glands. These glands make sweat, oil, and wax that you might find on your skin.

GOING VIRAL AT BODYLAND

Ah…ah…choo!

Oh no! A sneeze! Something is gumming up the works at Bodyland! We need to put a stop to it before the whole park shuts down! Call in the cleanup crew, aka your immune system!

Your immune system fights infections and diseases in the body. We're talking nasty

Historical Hoax

In 1869, it was reported that an ancient, petrified man, more than ten feet tall, had been unearthed in Cardiff, New York. The people who found it said it was proof that humans of giant proportions had once walked the earth. In reality, it was a massive scam orchestrated by George Hull. He hired sculptors to make this massive man-shaped statue, buried it on his cousin's farmland, and then hired two workers to build a well on the exact spot where he had buried the statue. Once it was unearthed, they started charging admission to see it and made A LOT of money. Scientists quickly started arguing it was a fraud, and Hull admitted to the hoax. But in the meantime, showman P. T. Barnum created his own fake giant and started charging admission. According to legend, one of the owners of the Cardiff Giant called out P. T. Barnum for fooling people (while still insisting the Cardiff Giant was real), saying, "There's a sucker born every minute."

stuff like bad bacteria, cold and flu viruses, and even fungi. Your white blood cells are a really important part of this system. There are several kinds of white blood cells, including lymphocytes. Lymphocytes are like the janitors of Bodyland, keeping things nice and tidy.

When your body first encounters a germ, it can take days for your immune system to find and tag all the virus clutter with antibodies. During that time, you'll get pretty sick. That can be dangerous—even deadly—depending on the virus.

It would be much better if your immune system knew what to tag for disposal the minute an intruder entered your body, before it can multiply and spread.

That's where vaccines come in. Vaccines are full of weakened or dead viruses. Stuff that can't really make you sick, but *can* get your white blood cells to start making antibody tags.

> **Vaccine:** a treatment that provides protection from a specific illness by exposing your immune system to a less dangerous version of the virus or bacteria that causes the illness

Those antibodies stay in your system long after the virus is gone. So if your cells see a virus that you've encountered before—whether through a vaccine or from getting sick—your immune cells will remember that virus and tag it immediately for disposal. Bye-bye, virus!

> My immune system does a great job cleaning up my body. Too bad it can't clean my room too.

Allergy Attack! When Your Body Overreacts

The immune system is great because it keeps out foreign invaders that could harm you. But sometimes your immune system will kick into high-gear invader-fighting mode for no good reason. When this happens, you have an allergy attack.

Let's say you inhale some plant pollen. To your body, this is a foreign invader—but not a super dangerous one like a virus or bacteria. So your body might create antibodies called immunoglobulin G—or IgG, for short—to handle things. It mostly just ignores the pollen. It's like, "Meh, pollen, who cares?"

Sometimes, though, your body will produce antibodies called immunoglobulin E, or IgE. These antibodies have the opposite reaction to pollen. They're like, "AHHHHH! POLLEN! THIS IS THE WORST THING EVER!!!! RUN!!!!!" These panicking IgEs made by your lymphocytes will roam around the body until they eventually come to rest on mast cells. Mast cells have special receptors that are a perfect fit for IgEs, and that's why

WHY DO SOME PEOPLE HAVE ALLERGIES AND OTHER PEOPLE DON'T?

Scientists still aren't sure why some people have allergies and other people don't. It's partially caused by your genes, and some of it is likely caused by your surroundings. Scientists think it's important to train your immune system early by exposing it to a range of friendly microbes and a wide variety of foods. This will help your body learn what to tolerate and what to attack.

Sometimes children will grow out of having allergies, and sometimes people will develop allergies as an adult. But again, we're not exactly sure why this happens. There's still a lot left to learn about these body overreactions.

they love to pair up. All is calm…until the next time your body encounters the allergen. This time, the IgE antibodies are sitting on a mast cell, and when they encounter the allergen, an allergic reaction begins.

You see, mast cells are full of a chemical called histamine, and when the allergen connects with the IgE sitting on a mast cell, it triggers a reaction in the mast cell that releases these histamines.

PRONOUNCER
Histamine = HISS-tuh-meen

Histamines put the ah-choo in allergies. They make you itch, make your nose run, all that annoying stuff. In some cases, this reaction can even cause swelling of the throat or make you throw up. Whether you're allergic to pollen, peanuts, mold, or anything else, this is the way allergic reactions work.

MYSTERY PHOTO

Focus your eyes on this mystery photo. Can you guess what it is? Turn to the next page for the answer.

It's pollen! Different flowers have pollen grains of all sorts of shapes and sizes. The spikes, nubs, and crannies help them hitch a ride on pollinators like birds and bees. Too bad these cool little grains make some of us a sneezy, wheezy mess!

OH, THE NERVE!

Are you ready for the ultimate speed ride? Well, then, you've come to the right place, my friend. Bodyland is home to a ride that goes more than 250 miles per hour—that's faster than a NASCAR racer! Make sure your seat belt is secure as you zip through the nervous system.

The nervous system is your body's way of communicating with itself. So when you're over at your friend's house and Señor Sassypants, their sheep, wants you to pet him, your nervous system sends a message to your arms and hands letting you bury your fingers in that curly coat of BAH-utiful fluff. But after Señor Sassypants has been rolling around in the mud and tries to jump in your lap, it's your nervous system that makes you leap back without even thinking about it. This leap is a natural reaction. Even before you can understand that you or your nice, clean jeans might be in danger, the body is already taking action to protect itself.

You can think of the nervous system like a bunch of extension cords running into a single power strip that plugs into the wall. The extension cords are the nerves running through your body, the power strip is your

SYSTEMS WITHIN SYSTEMS

The two main parts of the nervous system are the central nervous system and the peripheral nervous system.

- The central nervous system is made up of the spinal cord and brain—this is where the fastest messages get sent.

- The peripheral nervous system is made of nerves all over the body that eventually make their way to the central nervous system.

Fact-astic Voyage

Right now, you are reading these words, and nerves in your eyes are firing off signals to your brain telling it what you're seeing. This is happening at an astonishing rate, about 10 million pieces of information per second.

spinal cord, and the wall plug is your brain. Extension cords and plugs make an especially good analogy here, because messages transmitted throughout the nervous system are powered by actual electricity. Yes, your body is electric. (See page 95 in the brain section.)

The peripheral nervous system can also be broken down even further, into the voluntary and involuntary nervous systems. The voluntary nervous system controls things you want to do. So if you want to listen to your favorite podcast, the voluntary nervous system helps you grab a phone, pop in some earbuds, and push Play. The involuntary nervous system controls the stuff you never think about, like beating your heart, digesting your food, or moving that diaphragm muscle to keep your lungs breathing.

Your Nerves Are Painfully Aware

Let's say you're at the end of an amazing day of camping. There was that three-mile hike, where you're about 93 percent sure you saw Bigfoot. Then came the canoe trip around the lake that ended with a nice swim. Now you're sitting around the campfire thinking about how many more s'mores you can eat before going to bed. You bite into that gooey mess topped with chocolate and graham cracker, and then…OUCH! That marshmallow was practically still on fire!

What just happened here was a pain response. The instant the nerves in your tongue touched the molten marshmallow, they sent a signal to the brain. The brain immediately responded with a message like, "Get your mouth away from that fire slime NOW!"

There are special receptors for pain throughout your body called nociceptors. You can find them all over, including in your skin, muscle, bones, and stomach. They sense pain and send that information to your brain. And the brain "feels" this pain by receiving all these signals— even though there are no nociceptors in the brain itself! That's why people can have brain surgery while awake and not feel pain from it.

SUPER COOL SCIENTIST
DR. ROLAND ENNOS

Dr. Roland Ennos is a professor of biomechanics at the University of Hull, which means he looks at how living things are built and how they work. He's studied a huge variety of topics, from the design of tree roots and insect wings to the design of fingernails and fingerprints. He and his team found that fingernails have special reinforcements inside that make sure that if they rip or break, they break across the nail, and not down toward your nail bed (which would really hurt!).

PRONOUNCER
Nociceptor =
NOE-sih-sep-tur

MEGA MATCHUP
FINGERNAILS VS. TEETH

It's time for a battle of the body parts! These two hearty, hard helpers are both alive and dead at the same time, both awesome and kind of gross. In one corner, we have ten pickers, scratchers, and peelers: fingernails! And in the other corner, we have a set of grinders, gnashers, and chompers: teeth!

TEAM FINGERNAILS

- Your nails may be thin, but they are superstrong! They're made of keratin, the same protein that makes up hair, claws, feathers, and hooves.

- Your nails will just keep growing and growing if you let them! Lee Redmond was able to grow her nails out for thirty years without cutting them, and they grew to be more than two feet long! She holds the women's record for longest nails.

- Humans aren't the only ones with fingernails—all primates have them! Their nails probably break off from being used, but gorillas, chimpanzees, and other primates have been observed biting their nails. Just like us!

- Even though they're mostly dead, nails can provide clues about what's going on in the rest of your body. If fingernails change appearance—like shape or color—it can mean there are problems in internal organs like the lungs, kidneys, or liver.

- Nails can also be like tiny canvases! Humans have been decorating their nails for thousands of years. In ancient Babylonia, men would use kohl to color their fingernails, and in South America, Inca people would use sharp sticks and dye to draw small pictures on their nails.

TEAM TEETH

- The teeth in your mouth are like an elite food-eating squad with specialized tasks. The incisors, the big teeth in the front, are great for biting and cutting. The canines, those pointy side teeth, are for gripping and tearing. Your molars and bicuspids, the teeth with flat tops, are great for mashing and grinding.

- There's so much coolness to a tooth that you can't see! The root goes deep beneath the gums. The enamel on the outside of your teeth is dead and won't regenerate, but inside are living nerves and blood vessels.

- Your teeth are home to all sorts of friendly microorganisms, including lots of kinds of bacteria. These little fellas help you break down food and keep bad bacteria away. In return, your teeth provide a nice place for them to hang out and feast on the nutrients they need.

- A tooth's surface is harder than steel, and your molars can bite with 200 pounds of pressure. But even though they're strong, they are brittle and can be chipped.

- Humans have used artificial chompers, called dentures, for centuries. The earliest sets of replacement teeth were made of animal or human teeth, and wood, lead, gold, rubber, and porcelain were later used. Today, dentures are usually made of metal and plastic.

Which body part is cooler: Fingernails or Teeth?

YOUR VERDICT

YOUR BRAIN IS MAGIC

Sometimes zombies are right. They are obsessed with brains, and honestly, we get it. Brains are amazing. They let us see the world, they control memories, and they even give us feelings. But we prefer to use our brains rather than eat them. Sorry, zombies.

Brains?

HOW DOES YOUR BRAIN SEND SIGNALS?

Your brain does all this by receiving information and sending out signals at lightning-fast speeds through your nervous system. This happens in three main ways:

1. **Taking in signals from the five senses (touch, taste, sight, sound, and smell).**
2. **Passing signals from one part of the brain to other parts of the brain.**
3. **Sending signals out to the rest of the body, like your legs, heart, or mouth.**

So when you hear a "meow," your ears will take that sound and send it to the brain. The brain will decode it so you know that you are hearing a cat. From there, the signal might trigger your memories, which let you know that it's not just any cat meowing—it's your long-lost cat, Captain Hairballs, back from years of adventure on the high seas! Then the signal might pass to the part of the brain that feels emotions, so your eyes fill with tears at seeing your old friend. And finally, your brain might send a signal back to your body, telling it to scoop up Captain Hairballs and shower her with kisses. All in a fraction of a second!

Your Brain Is Electric!

Even though you don't have to plug yourself in, you still run on electricity. That's because cells in your body send signals using tiny electric pulses. This is especially true for cells in the brain called neurons. They sort of look like trees, with lots of spindly roots at one end, a long skinny trunk in the middle, and then branches at the opposite end. All your thoughts, memories, hopes, and dreams are held in these special little cells. Mind-blowing, right? I mean, neuron-blowing, right?

PRONOUNCER
Neuron = NIR-ahn

Fact-astic Voyage

Your brain has anywhere from 80 BILLION to 100 BILLION neurons. Which is A LOT. They aren't all active at the same time, but while you are awake, there are enough neurons zapping signals to one another to make about ten watts of electricity. That could power a small light bulb!

When these cells are activated, an electric charge runs through the long trunk part. *ZAP!* When that charge reaches the branches at the very end of the cell, it triggers them to shoot out a bunch of little chemicals called neurotransmitters. *SQUIRT!* Those chemicals float away from the cell over to other nearby cells, where they attach to tiny receptors on the roots, like a key fitting into a lock. *CLICK!* If enough neurotransmitter chemicals click into place on that cell, it will also activate and send its own electric impulse (*ZAP!*). That will cause it to release more chemicals (*SQUIRT*), which will activate more cells (*CLICK*), and on and on in a long chain reaction. This is how neurons send messages. It's *ZAP, SQUIRT, CLICK* all throughout your brain!

So if I stick this cable up my nose, can I power my cell phone?

Historical Hoax

Are robot brains superior to human ones? In 1770, Wolfgang von Kempelen built a chess-playing robot called "the Turk" to impress the empress of Austria. This remarkable machine was taken around Europe and North America to demonstrate its incredible chess-playing abilities, defeating many humans including Napoleon Bonaparte and Benjamin Franklin. How was this amazing robot able to do this years before the invention of computers or even typewriters? Well, it wasn't. Inside the machine, there was an actual human who would control the Turk and play the game on its behalf. In 1912, Spanish scientist Leonardo Torres y Quevedo did invent an actual chess-playing robot called El Ajedrecista, which successfully played against—and beat—human opponents! You could argue that this bot was the very first computer game.

HOW DOES MEMORY WORK?

Memory is this totally amazing thing. It's like a scrapbook you carry with you always, or like time travel for your brain. You simply think of the past, and memory brings back sights, sounds, smells, and emotions from long ago. Not even the latest smartphone can do that! So how do you make a memory?

A RECIPE FOR MAKING MEMORIES

Ingredients:
- 1 brain
- a very large number of neurons
- 1 memorable event
- parsley (optional)

1. Start by mixing some very memorable elements, like a heat wave, a beach day, and a crab that you accidentally stepped on. OUCH!

2. A lot of neurons in your brain will activate, or "light up," when you step on that crab.

3. Your brain will find those active neurons and tag them so they are part of a crab-stepping memory.

4. That crab-stepping memory is now ready! Let it cool down, then have your brain move that freshly baked memory from short-term storage into long-term storage.

5. Memories keep very well, so no need to freeze them. Whenever you want a helping, just think about this incident or do something that will remind you of it—like visiting the same beach again. Many of the same neurons from the original crab-stepping incident will light up once more, so it's sort of like you are living it a second time. OUCH ALL OVER AGAIN!

6. Garnish with parsley. Bon appétit!

Déjà Vu: Didn't This Happen Already?

You're with your friend, who's telling you a story. Suddenly, time seems weird, almost like you know what your friend is about to say next. It could be that your friend repeats the same stories… or it could be déjà vu!

Déjà vu is a French term that translates as "already seen." It's as mysterious as it is French. Some scientists think we have déjà vu when incoming memories go straight from our senses to long-term storage, skipping over our short-term memory storage. This could give us the sense that we're remembering something that is still happening. Other scientists think déjà vu occurs when the part of the brain that marks things as familiar gets mixed up and tells us we recognize something that is actually new to us. Or maybe déjà vu happens when we experience something familiar, but we forget where we first saw or heard it—was it in a book or on TV? Scientists are still trying to figure out this brain glitch that is as mysterious as it is French.

> Is it just me, or does that last line sound familiar?

BRAIN GLITCHES GALORE!

Déjà vu isn't the only brain glitch out there. Psychologists have studied lots of them, and of course they all have very French names.

1. Jamais vu

- Translation: Never seen
- This is the feeling that something familiar is actually brand-new to you.

2. Déjà entendu

- Translation: Already heard
- When you think you've heard something before.

3. Presque vu

- Translation: Almost seen
- The feeling that you are about to have a big breakthrough idea, but it doesn't come.

4. Déjà rêvé

- Translation: Already dreamed
- When you have trouble telling the difference between dreams, memory, and reality.

SUPER COOL SCIENTIST

DR. ANNE CLEARY

It's tough to study déjà vu because you never know when the feeling will strike. But Dr. Anne Cleary found a way to trick people into experiencing it, using a video game. She's a cognitive psychologist at Colorado State University, where she has people use virtual reality goggles to tour scenes created using a game called *The Sims*. She shows them a series of rooms, and some of the rooms have the same layout. When the players see a room similar to one they already saw but that they forgot seeing, they often get the feeling of déjà vu. She hopes this research can shed light on this common but puzzling phenomenon.

EYES ARE THE WINDOWS TO THE BRAIN

Your brain and your eyes are best buddies. They're always around each other, they're constantly passing messages back and forth, and they're even directly linked thanks to something called the optic nerve. This cable-like bundle of nerves allows your eyeballs to send images to your brain almost instantly. Your brain then decodes these images so you know what you are looking at. We see the world thanks to the teamwork between these two fast friends.

Vision starts when light from the world goes to the back of the eye. Once there, the light is absorbed by a bunch of tiny cells that turn that light into signals for the brain to make sense of. There are two kinds of light-absorbing cells: rod cells and cone cells. Each kind has different powers.

Rods, rods, we're so fun! In human eyes, we're number one!

Cones, cones, we're so cool! We sense color, and color rules!

Oh yeah? Without us, they wouldn't be able to see at night or in dark rooms! We work great in low light. Cones don't!

Sure, but rods see only blurry images. Cones show fine details. Plus, there are three of us! Cones that take in green light, cones that take in red, and cones that take in blue! Add us three together, and you can make all the colors of the rainbow!

Whatever. Three's a crowd. Rods forever!

RODS

CONES

ROD

GO! CONES

CONES

Do We All See the Same Colors?

We all know what blue is, right? It's the color of the sky, or blue jays, or Cookie Monster. But what if the color you see as blue looks orange in my brain? So I see an orange sky, orange blue jays, and an orange Cookie Monster. But since everyone calls that blue, I just think orange is blue. *Could we all be seeing different colors and not even know it?*

The idea that none of us see the same colors gives me the blues...or the oranges? I don't even know anymore!

A META MOMENT OF EWW

Have you ever wondered why we say, "Eww!" when we encounter something gross? Turns out that it goes back all the way to why we experience disgust in the first place. Disgust is all about preventing something bad in the outside world from getting inside our bodies. The easiest way something can get into our insides is through our mouth. If you look in the mirror while making the noise "Eww!" you can see that it's basically a way of closing your mouth, or if you had something in your mouth already, it's the perfect shape for spitting it out. Same thing goes for other reactions of disgust, like sticking your tongue out or saying, "Blech."

The short answer is: We don't know! Every brain is unique and sees the world a little differently. We *do* know that everyone's eyes work pretty much the same. They see color by absorbing light waves. Each color is a different size of wave—which means each color has a different *wavelength*. The colors go on a spectrum, from those with the shortest wavelengths (like blue and purple) to the colors with the longest wavelengths (like orange and red).

The light wave reflected off a blue crayon is the same length no matter who is looking at it. So in a sense, we're all seeing the same colors. The question then becomes, do our brains process that color the same way? Maybe some brains see that wavelength and make it look orange, while others make it look blue. Brains are weird, so this is possible.

To make things even weirder, some people are able to see more greens than other people. Remember those light-detecting cone cells we talked about? Women sometimes have more than one kind of the green-detecting cones in their eyes. They may have ones that sense normal green and also ones that sense more bluish green. This makes them super seers. So it's totally possible that no two people see the world exactly the same.

WHEN SEEING ISN'T BELIEVING

The brain is busy. It's got to think your thoughts, control your body, process incoming information, and DO EVERYTHING to keep you alive. So sometimes the brain takes shortcuts in how it processes information. Optical illusions often take advantage of these shortcuts to make mind-bending tricks of the eye. Check it out!

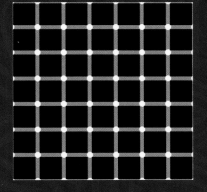

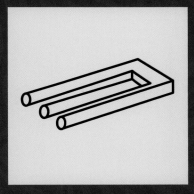

When you move your eyes around this image, it should look like the circles are spinning! The illusion seems to come from how we process patterns at the edges of our vision. Feeling dizzy?

You might see black dots appear in the white circles of this grid, but the black dots disappear when you look right at them. Weird!

If you look at the left side of this, it clearly has three rods. But when you look at the right side, there are only two! Brains have trouble with images like this.

CAN WE CONTROL OUR DREAMS?

We all dream four to six times a night, but we usually forget most of it. The dreams come in cycles and get longer as we snooze. The first dream of the night might be just twenty minutes long, but the last one could run almost an hour! And even though it feels like rest, some parts of your brain are actually super active while you dream.

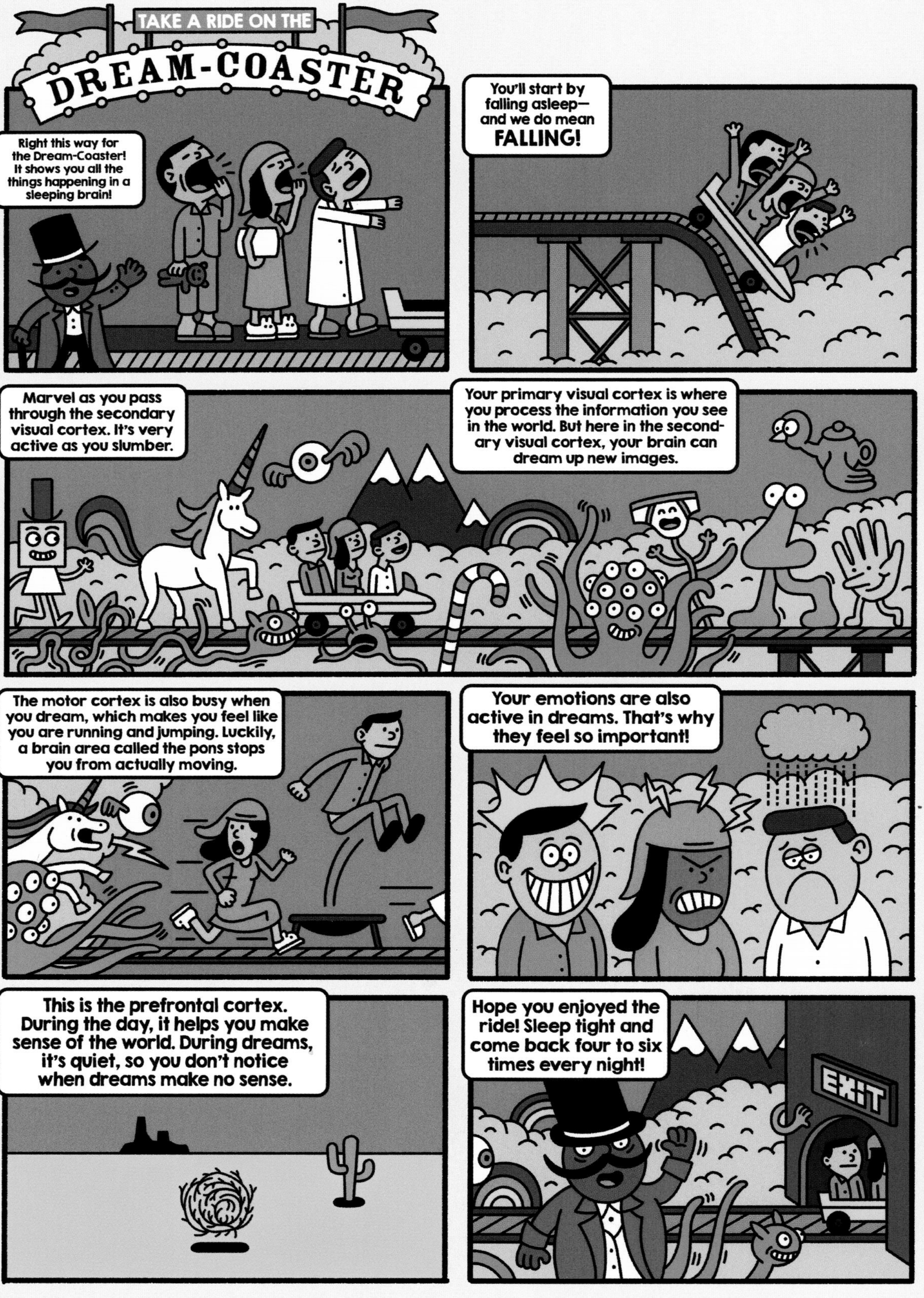
TAKE A RIDE ON THE
DREAM-COASTER
Right this way for the Dream-Coaster! It shows you all the things happening in a sleeping brain!
You'll start by falling asleep—and we do mean FALLING!
Marvel as you pass through the secondary visual cortex. It's very active as you slumber.
Your primary visual cortex is where you process the information you see in the world. But here in the secondary visual cortex, your brain can dream up new images.
The motor cortex is also busy when you dream, which makes you feel like you are running and jumping. Luckily, a brain area called the pons stops you from actually moving.
Your emotions are also active in dreams. That's why they feel so important!
This is the prefrontal cortex. During the day, it helps you make sense of the world. During dreams, it's quiet, so you don't notice when dreams make no sense.
Hope you enjoyed the ride! Sleep tight and come back four to six times every night!
EXIT

Why Do We Dream?

Imagine doing something every day of your life but having no clue why you do it. That's what it's like with dreams. We all dream, but scientists are still trying to figure out why. Are dreams your brain processing memories and images from the day? Or are dreams just fun stories your brain makes up when it's bored? Over the years, scientists have come up with lots of possible explanations.

Garbage Disposal Theory

This theory says that when we dream, we are watching our brains throw out useless memories and information. After all, you experience tons of things every day that you quickly forget, like what you ate for breakfast or the color of your friend's shirt. The problem with the Garbage Disposal Theory is that research into dreams shows that, if anything, dreams are full of stuff we are *trying* to remember. That's why you might dream about a book you just read or a video game you are trying to beat. So we can pretty much toss out the Garbage Disposal Theory.

Threat Simulation Theory

According to this theory, our brains use dreams to help us practice facing threats. So when you're running from a jaguar in a dream—or showing up unprepared for a test—that's just your brain practicing how it might handle that scary situation. Then, if either of those things happens in real life, you might be better prepared. The problem with this idea is that not all dreams are scary or threatening. In fact, some are really fun—like when you dream you can fly or that pizza fountains exist.

Fact-astic Voyage

Dreams are great for inspiration. Physicist Albert Einstein said the early beginnings of his famous Theory of Relativity came from a dream. Artist Salvador Dalí would dream of fantastical images, then intentionally wake himself up to paint them. Musicians Paul McCartney from the Beatles and pop diva Taylor Swift both say they wrote songs after hearing something in a dream. So go to sleep, and you might have your next big idea!

Wish Fulfillment Theory

This theory is kind of like the opposite of the Threat Simulation Theory. It says when we dream, we are doing things that we'd like to do in real life but just can't for some reason. So in dreams, you might meet your favorite athlete or be able to breathe underwater. But again, not all dreams are wishes come true. A lot of them are stressful, boring, or just weird. This theory seems like wishful thinking.

Mood Regulator Theory

In this theory, dreams help you deal with difficult moods or feelings. Say you didn't get invited to a sleepover, and you're super bummed about it. When you sleep that night (at your own house, sad emoji), you'll still feel down. But according to this theory, in your dreams you might feel other things too, like confidence or joy. And you might revisit happier memories. When you wake, you'll feel more upbeat and ready for a fresh start. There's some research to support this theory, but it's hard to fully prove since we can't see all the things happening when we dream.

Your Theory Here

Scientists are still looking for a good explanation of why we dream, so why not come up with your own? Remember a few of your dreams, write them down, and think about why you might have had those dreams. Do they have anything in common? Are they related in any way? Once you have a theory, give it a name and start telling the world! Maybe one day, your theory will win a Nobel Prize! Hey, you can always dream, right?

SUPER COOL SCIENTIST

DR. SAMER HATTAR

Without enough sleep, we're more likely to get sick, we're crankier, and our reflexes slow down. Not good! Dr. Samer Hattar wants to help improve our sleep health! He works at the National Institutes of Health and discovered special cells in the eyes that absorb light and send a message to the brain to let us know when it's daytime. These cells specifically absorb blue light from the sun. But it turns out computer screens also give off blue light, tricking those cells into sending signals that it's day even after dark. This confuses your brain and makes it hard to fall asleep. So Samer recommends putting away all screens at least two hours before bed.

DESIGN YOUR OWN DREAM!

It's possible to control your dreams, but it isn't easy. Follow these steps to see if you can build your ideal dream.

- Think of something you want to dream about, like vacationing on Mars.
- Throughout the day, remind yourself you want to dream about Mars.
- Just as you're falling asleep, tell yourself, "I'm going to dream about Mars," over and over.
- As you are falling asleep, picture the dream you want in your head.
- Keep a pen and paper nearby. When you wake, write down what you remember of your dream.
- Keep trying! It can take a few days, but hey, hiking the mountains of Mars will be worth it, right?

WHERE DO FEELINGS COME FROM?

Next time you get a Valentine's Day card, it better have a brain on it. Because hearts don't make feelings like love or friendship—our brains do! Other body parts play a role as well—like our guts (they would make a good Valentine's Day decoration too). But when it comes to our emotions, from happiness and sadness to anger and anxiety, the brain is key. So stop taking all the credit, hearts! Stick to pumping blood!

Happiness! 😀

You found a dollar! You aced the test! You hugged a friend! You high-fived an astronaut! There are lots of ways to get that glowing, uplifting sensation we call happiness. Sometimes it comes easily, like when we eat ice cream (instant joy, am I right?). Sometimes it comes after lots of hard work. But no matter how you get there, the feeling is brought to you by chemicals in your brain called neurotransmitters.

Swoosh! You shot a perfect three-pointer and now your basketball team is in the lead. Your brain suddenly fills up with a neurotransmitter called dopamine. This lets the rest of your brain know, "Hey, that felt great!" Now you're excited and motivated. You can't wait to score another point.

You look to the bench and see all your teammates cheering you on, so your brain releases another chemical called oxytocin. Oxytocin makes you feel closer to your friends. At the same time, you get more chemical signals from serotonin, which will keep you feeling good long after the game is over. Another important player on team happiness is endorphins. Endorphins can actually dull pain, so you might not notice that your knee is a little sore from all that running.

This mix of brain chemicals helps create feelings like happiness and contentment. No two people are exactly the same, so everyone's mix is a little different. Some people feel happy pretty easily. For others, it can take a little more to make them smile. When your mix of these chemicals is high, you feel good and you're less stressed out. They even help boost your immune system to keep you from getting sick. No wonder happiness makes us so…you know, happy!

PRONOUNCER
Oxytocin =
ox-see-TOE-sin

Serotonin =
sair-uh-TOE-nin

Sadness

If you've ever felt sad, you are not alone. It's not just other people. Lots of animals seem to feel blue from time to time. In fact, one way to help us understand why we might feel sad is to look at the role this mood plays in animals, especially social animals. Social animals, like wolves, fish, and us humans, live in groups. Groups help protect individuals. They can work together to find food, share shelter, and even raise kids. It's a really great survival strategy.

But these social relationships can also affect moods, for better or worse, because social animals can be very competitive. For instance, if two lobsters meet in a tank, they will sometimes compete to see which one gets to be the top lobster. When one lobster regularly wins, that lobster will become more dominant. Winning or losing causes chemical changes in the lobster's brain. The chemical changes in the losing lobster cause it to act in a way that looks like sadness to us. It might swim less or stay in the corner of the tank and hide from other lobsters.

Another reason social animals might feel sad is because we form strong bonds with others. That can lead to a feeling of loss when someone we love dies. So sadness and being social seem to be linked in some deep ways.

Is There an Upside to Feeling Down?

It's safe to say that sadness is a total bummer, but it might eventually lead to some good things. Some scientists think that after a while, sadness can actually motivate animals to make changes. Maybe they'll decide to form new bonds, or even work on improving their social standing.

One way to do that, of course, is by socializing. When a young songbird spends time with more experienced songbirds, the younger one will be better able to learn and practice its songs. Young fish who spend more time learning from others become more confident and better at spotting danger. Sadness might be a side effect of being social creatures, but sometimes it can help us become safer and happier. If you're feeling sad, remember there are a lot of people who can help you and want to help you—like your family or teachers. You don't have to deal with sadness on your own!

Anger 😠

You've waited all day. The clock strikes three, school lets out, and you rush home. You open the fridge expecting to find that half burrito you saved from last night's dinner—so chewy, so cheesy, so full of beans. But wait, it's not there! Just then your brother walks by, licking his fingers. "That was a dope burrito, dude," he says. And that's when you lose it. Commencing anger meltdown in 5…4…3…2…1…

Sadness isn't the only emotion that can be overwhelming. Anger is often super intense too.

When we're angry, it's usually because our brain sees something it thinks is a threat—like when someone bullies us or takes away something we want (like a burrito!). Threats kick our brain into high gear, getting us ready for something scientists call the fight, flight, or freeze reaction. That means you can:

- **Fight to defend yourself.**

- **Take flight by running away.**

- **Or freeze, perhaps to avoid drawing attention to yourself.**

> ## THINGS TO TRY WHEN YOU'RE FEELING SAD
>
> 1. Get some exercise! This can bust stress and boost your mood.
>
> 2. Pet an animal. It's relaxing!
>
> 3. Meditate. Try sitting calmly and taking several slow, deep breaths in a row to reset your mind and mood.
>
> 4. Talk about your sadness with a friend or a trusted adult.

When this happens, your brain sends a message to glands right behind your stomach to release two hormones—adrenaline and cortisol—that let your body know it's GO time. When adrenaline and cortisol are pumping through your body, you can run faster and farther, plus you have more fuel for your muscles and brain.

All that can happen before you even realize there's a threat! Next thing you know, your face is red, your heart is racing, your palms are sweating, and you might even feel like hitting something. Adrenaline and cortisol are powerful stuff—and if you were facing a threat like a hungry leopard, they just might help you survive. But most of the time, you aren't in serious danger, and these hormones just make you defensive, angry, and hard to be around.

Understanding why you're angry can help you deal with your feelings in a productive way without hurting people or smashing things. Take deep breaths, close your eyes, and go to your happy place. And remember, there will be other burritos. It's nothing to lose your cool over.

Anxiety

Butterflies in the stomach. Heart beating fast. Twitchy leg. Sweaty pits. Don't panic, it's just anxiety. Nervousness and anxiety are universal feelings. And like happiness, sadness, and anger, this feeling has its roots in human evolution.

Anxiety developed in our animal ancestors to shape their behavior so they'd survive and have kids of their own. Imagine if an animal *didn't* get anxious. That brave creature would be fine playing near tall cliffs, eating weird berries, and walking past hungry lions. That brave creature wouldn't last long. An anxious animal, on the other hand, would stay far from those deadly cliffs, pass on the probably poisonous berries, and scram if a lion was nearby. It would be more likely to live long and pass down its anxious ways to its kids, and those kids to *their* kids, and so on.

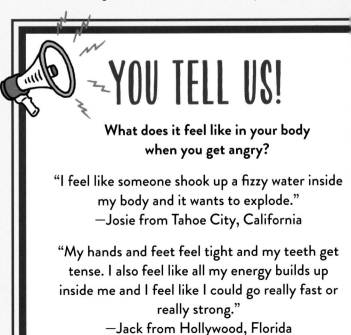

YOU TELL US!

What does it feel like in your body when you get angry?

"I feel like someone shook up a fizzy water inside my body and it wants to explode."
—Josie from Tahoe City, California

"My hands and feet feel tight and my teeth get tense. I also feel like all my energy builds up inside me and I feel like I could go really fast or really strong."
—Jack from Hollywood, Florida

That's how creatures like us end up with a built-in fear alarm in our brains and bodies. We call this alarm *anxiety*. Most of us don't worry about tall cliffs or hungry lions, but this alarm might keep us from playing near traffic or climbing too high on the monkey bars. But sometimes we feel anxious about things that don't threaten our physical health, like a math test or figuring out where to sit in the school cafeteria. It's that same old built-in fear alarm, but the situations have changed.

Why Do We Get Butterflies in Our Stomachs?

Emotions don't just live in our heads; they're in our bodies too. When you're in fight, flight, or freeze mode, your blood flow is diverted away from your stomach to your muscles in your arms and legs. That slows down your digestion, and the drop in blood flow to your gut gives you that butterfly feeling. It's a side effect of your body's getting ready for action.

If you're feeling overwhelmed by anxiety, try splashing cold water on your face or putting an ice pack over your eyes. This actually slows down your nerves a bit. You can even just sit quietly and take some deep, calming breaths. It's one way to say, "Thanks for keeping me safe, anxiety, but I don't need you right now."

YOU'RE SO CUTE I COULD EAT YOU UP!

Cute aggression—the desire to bite, squeeze, or eat something because it's so cute—is a common emotion. And this kind of emotion, which is similar to crying when you're happy or laughing when you're nervous, is called dimorphous expression. The same thing is happening when you react to something cute by saying, "Awwww," which usually is paired with an exaggerated frown. Seeing cute things isn't actually making you sad or angry. Some researchers think these kinds of dimorphous expressions may help us better regulate intense emotions, like those brought on by seeing an adorable puppy, or an adorable baby, or—better yet—a dog and a baby cuddling together while napping.

MEGA MATCHUP
READING VS. WATCHING

In this brain-based blowout, book lovers go head-to-head with film and TV buffs. In one corner, we have the curl-up-on-the-couch pleasure of the printed word: reading! And in the other corner, we have screens of all sizes made to entertain the brain: watching!

TEAM READING

- When you read, your brain is doing some amazing mental gymnastics. Since reading is not something we naturally figure out (like talking or walking), our brains have patched together a cool system for understanding the written word that involves many different parts of the brain. However, this is also why it's pretty common to have challenges in reading, like dyslexia.

- Reading forces you to use your imagination to fill in the visual blanks. This improves your brain's connection powers long after you put down the book.

- When you read, you're giving your brain a literal workout. When you read about someone running or jumping, the parts of your brain responsible for running or jumping are also activated.

- Want better focus and memory? Read a book! This helps get your brain firing on all cylinders and boosts your ability to store information.

- Relax and get some much-needed sleep with a book! Reading books as part of your bedtime routine can help you fall asleep more easily. It's much better than watching screens before bedtime, which can actually make it harder to fall asleep.

TEAM WATCHING

- When we watch with other people, it engages our mirror neurons. When we sense others having the same feelings, it makes our own stronger. There's even research suggesting that emotional movies can help people form stronger bonds.

- Watching something funny is a workout! Laughing—a good belly laugh for about fifteen minutes—has the same effect in lowering your blood pressure as exercising. Laughing can boost your immune system too!

- The music in a TV show or movie can make us feel things more strongly. Many studies have shown that music has a direct effect on our emotions—heightening fear, anger, happiness, or calm.

- When you're watching a movie and you see a ball fly toward the screen, you might flinch or even try to duck out of the way. That's because movies hack our senses, making us feel like we're part of the action. When was the last time a book made you want to duck?

- Rewatching your favorite movies can be a calming and relaxing experience. Because we've seen it before, it's easy for our brains to process, and since we know what's going to happen, there are no surprises. This relaxes our brains.

Which brain-stimulating activity is cooler: Reading or Watching?

YOUR VERDICT

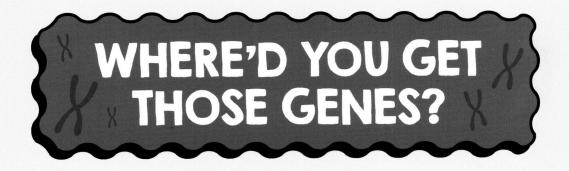

WHERE'D YOU GET THOSE GENES?

WHAT IS DNA?

Genes and jeans sound the same, but they are very different. Your jeans are a pair of denim pants that go well with pretty much any outfit. Your genes are a complex set of instructions in your cells that go well with being a living thing. You probably got your jeans from a store or online, but how did you get your genes? Before we find out, let's look at what exactly they are.

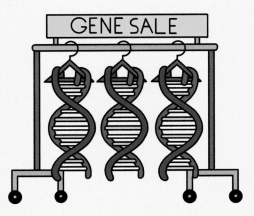

ZOOMING IN ON GENES with the Handy-Dandy Zoom-Ray

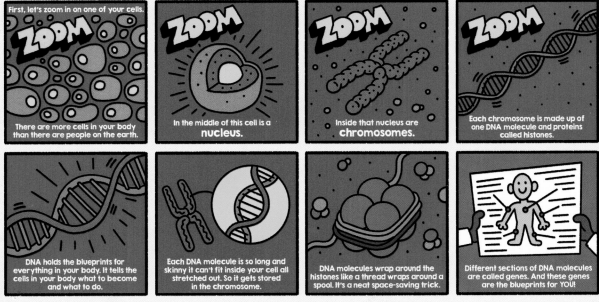

First, let's zoom in on one of your cells.

ZOOM

There are more cells in your body than there are people on the earth.

ZOOM

In the middle of this cell is a **nucleus.**

ZOOM

Inside that nucleus are **chromosomes.**

ZOOM

Each chromosome is made up of one DNA molecule and proteins called histones.

DNA holds the blueprints for everything in your body. It tells the cells in your body what to become and what to do.

Each DNA molecule is so long and skinny it can't fit inside your cell all stretched out. So it gets stored in the chromosome.

DNA molecules wrap around the histones like a thread wraps around a spool. It's a neat space-saving trick.

Different sections of DNA molecules are called genes. And these genes are the blueprints for YOU!

Deoxyribonucleic acid: Also called DNA, a molecule shaped like a double helix, or a twisted ladder. Each ladder rung is made up of a pair of compounds. The pattern of these compounds makes up the instructions that tell your body how to be and what to do.

WHY DO I LOOK THE WAY I DO?

When it comes to chromosomes, that genetic stuff dwelling in your cells, most people have forty-six of them. Half come from your biological dad, and the other twenty-three come from your biological mom. That means we have DNA and genes from both of our biological parents, and it explains why we have traits like theirs.

Families can be made in many different ways, so your biological parents may not be a part of your family. You might not even know your biological parents. But the DNA that makes you *you* came from somewhere—and it came from them!

Some traits have to do with how you look, like your freckles, your brown eyes, or

PRONOUNCER
Chromosome =
CROE-muh-sohm

Fact-astic Voyage

The X-Men and Ninja Turtles are mutants, sure—but so are you. In fact, we all are. Mutations are random changes that can happen in DNA. Some we inherit from our parents, and others can happen during our lives. Sometimes these mutations can cause disease, but other times they can have no real effect or even be helpful. Red hair is caused by a mutation, as is the ability to drink milk. Most mammals can't digest lactose (the sugar in milk) once they stop nursing, but some humans have developed a mutation that allows them to keep drinking milk well past babyhood. So maybe milk should have a new slogan: Milk is for mutants!

MYSTERY PHOTO

Focus your eyes on this mystery photo. Can you guess what it is? Turn to the next page for the answer.

your extra-big big toe. Others have to do with how your body works, like if you need glasses, sneeze when you look at the sun, or think cilantro tastes like soap. All those traits are in your genes.

But even though all our genes come from our biological parents' chromosomes, your chromosomes aren't identical to theirs. Here's why:

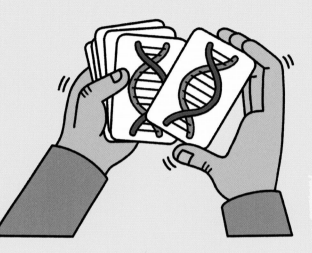

- **Each biological parent has two sets of chromosomes (just like you).**

- **Before the chromosomes are passed to you, these two sets get shuffled like a deck of cards.**

- **Then each biological parent passes down one-half of these shuffled-up chromosomes.**

So you end up with one set of chromosomes that's a mix of your bio dad's chromosomes, and one set of chromosomes that's a mix of your bio mom's chromosomes.

That's why we're like our biological parents, but still unique! It's complicated and also awesome.

ANSWER!

It's ice cream! If you can eat ice cream, you have a genetic mutation to thank! Ice cream is made from milk, and the inability to consume milk is called lactose intolerance. Lactose is the sugar in dairy that's hard for some people to digest. People who can eat dairy make an enzyme in their guts called lactase that helps them break it down. All baby mammals make lactase, but most stop making it when they stop drinking their mother's milk. Some humans keep making it, though!

READING THE BOOK OF YOU

In your teeny-tiny microscopic cells is a ton of information. The DNA that is carried in almost every cell of your body includes all the instructions for how to make you. And every human has somewhere between 20,000 and 25,000 genes carried in that DNA.

All that information is called your genome. And your genome makes you who you are.

Scientists first started planning to map all the DNA in one human genome in 1988, and it took them about fifteen years to complete that task. Now, thanks to high-powered computers, an entire human genome can be mapped out in minutes.

> **Genome:** the complete set of genes that makes up DNA; all the genetic information for an organism

MOMENT OF EWW

BLOOD, GUTS, AND DNA

DNA is present in almost every cell in your body, with just a few exceptions: red blood cells and hair. That means that you could map your genome with a single cell from any body part, besides those. You could use hair if the follicle—that little nub at the bottom of a hair strand—is still attached. You can't use red blood cells, but white blood cells have DNA. Saliva, nasal mucus, urine, and vomit usually will contain skin or tissue cells with DNA, so you could probably use those too, if you really wanted.

We know that two of the genes that affect eye color are on chromosome 15, and the gene that controls if you can drink milk or if you're lactose intolerant is on chromosome 2. But there's still so much we don't know.

Now, let's say each person's genome is a book. No two books are identical, but if you compared your book with another human's, they would be 99.9 percent the same. That's a lot of similarity, right? But in that 0.1 percent that is different are all the things that make you genetically unique. And that uniqueness comes from your ancestors. The chromosomes that were passed down to you from your biological parents were passed down to them from their biological parents, and so on. There are stories hidden there, and scientists are learning how to decode them.

SUPER COOL SCIENTIST

DR. JANINA JEFF

Dr. Janina Jeff is a geneticist. She's working to understand all the information that's stored in our genes. She's particularly interested in looking at the information stored in the genes of people of African descent. Not only are these the oldest genomes, but looking closely at them will help develop medicines that work well for people of all backgrounds.

WHY YOU CAN'T USE DINO DNA TO MAKE A NEW DINO

Scientists have been able to successfully clone some animals, which means making an exact replica of an animal, by using their DNA. The first was Dolly the sheep, who was cloned in 1996.

But what if you want to build an animal that isn't alive anymore, like something that went extinct? You might be able to find a bit of DNA in a fossilized bone, but the problem is that DNA decays, or breaks down over time, and is totally unusable after 1.5 million years.

Alas, dinosaurs went extinct 65 million years ago. So even if we found a sample of dino DNA, the instructions would be so old and ratty that even the best builders couldn't use them to build a dinosaur.

Wait. So *Jurassic Park* is a lie? I feel so cheated.

MOMENTS OF UM

HOW DOES COFFEE KEEP YOU AWAKE?

Coffee has a chemical called caffeine, and that's what keeps you awake. Here's how it works: Caffeine is a twin to another chemical called adenosine. Over the course of a day, your body makes adenosine, and it locks into cuplike spots on your cells, making you tired. Since caffeine is a twin to adenosine, it fits in those same cell receptors, but it doesn't make you tired. So when adenosine comes around, looking for cells to sit on, it finds all the cells are already full with caffeine, so it can't make you tired.

WHY ARE BRUISES PURPLE AND BLUE, AND HOW DO WE GET THEM?

When light passes through your skin and other tissues, it makes the blood underneath look blue. Blood flows through tiny tubes under your skin, and when you bump your knee or your elbow, you are breaking these tiny little tubes. Since your skin isn't broken, the blood has nowhere to go but right underneath your skin, making a bruise. Bruises will look different on different people's skin, but we all see a bluish or purplish color in the beginning, even though the blood underneath is still red.

WHAT IS GOING ON INSIDE YOUR BODY WHEN YOU CRACK YOUR KNUCKLES?

That cracking noise you hear is a bubble being formed! Joints are places in your body where two bones come together, like your knees or knuckles. There is negative pressure in these joints—like a tiny sucking vacuum—and that's what holds the bones together. When you pull the joint, the pressure changes and nitrogen gas that's usually in your bones is pulled out, forming a bubble that quickly pops.

WHY DO YOU GET DIZZY WHEN YOU SPIN?

Your inner ear has a special organ that helps it sense tilt, spin, and changes in speed. It's called the vestibular complex. It's a series of tiny tubes lined with microscopic hairs and filled with ear gel. When you move, the ear gel sloshes around in these tubes. Those tiny hairs in the tubes sense that sloshing and send signals to your brain telling it about the movement. After you stop spinning, it takes a bit for the gel to stop moving too. That sloshing in your ear makes you feel like you're still moving and creates that dizzy sensation.

WHAT CAUSES HICCUPS?

Each hiccup starts with a signal from your diaphragm, a flat muscle under your lungs. When you have the hiccups, your diaphragm contracts quickly, causing you to make a very sudden, quick intake of air. This causes your glottis to close suddenly, which is where the "hic" sound comes from. The glottis sits atop the trachea, the passageway where air travels in and out of your lungs.

WHY DO WE HAVE FINGERPRINTS?

We're still not really sure, but there are a few different ideas. One is that they help make our fingers and hands more sensitive. Another idea is that they act as strengthening rods that prevent us from getting blisters. If you've gotten blisters, you'll notice you don't get them in the areas of your hands and feet where there are prints.

WHY CAN'T YOU TICKLE YOURSELF?

Your skin is covered with receptors, and it's their job to tell your brain and spinal cord what you are feeling on your skin. In many ways, it's more important for your brain to keep track of the things you can't control—like people, animals, or objects that brush up against your skin. When you do something to yourself like tickling, your brain dulls the feeling because it knows you are doing it. You can still feel it, but it's not as intense.

PART 4
MICRO-VERSE

IT'S A (VERY) SMALL WORLD AFTER ALL

Psst. We want to let you in on a little secret: No matter where you are right now, you are not alone. In fact, you have *a lot* of company. There are trillions and trillions of tiny things all around you. Things so small you can't see them without a microscope. In the air! Underground! In your food! On your body! And they're *ALIVE*!

This secret microworld is a lot like ours: crowded, busy, and surprisingly gassy (more on that in a bit). Most of these eensy-teensy organisms are harmless, and all of them are super fascinating. We call them microbes. Let's meet a few.

- **Bacteria:** These microscopic living things are made up of just a single cell. There are many different kinds of bacteria, and they can be found pretty much everywhere on earth, from mountaintops to the ocean floor. Bacteria are epic eaters too. Depending on the type, they can feast on anything from old food to hard rocks.

- **Fungus:** We used to think these were plants, but now we know they are their own special kind of life. They come in many shapes and sizes. They digest their food externally, meaning they break down stuff around them and then absorb the nutrients. Fungi include mushrooms, which can be seen with the naked eye, but many are teeny-tiny—like the yeast in bread!

- **Microfauna:** These tiny animals can't be seen with the naked eye. They include the incredibly tough tardigrades, slithering nematodes, and itsy-bitsy mites—the micro-sized cousins of ticks and spiders.

Microbe: a living thing so small it can be seen only with a microscope

The following are speech bubbles from the illustration at the top of the page:

Are you interested in getting to know these little critters better? Great!

We'll just need to turn the Zoom Ray on. *VWOOSH!*

And set it to Shrink Mode. *TICK TICK TICK BING!*

And then press the Shrink button. 3, 2...hold on. Just want to make sure you're ready.

You're sure?

Okay! Here we go! 3, 2, 1...Shrink Mode! *ZHWUH-VOOOOOOM!*

MEET YOUR MICROBES!

Wow. It worked. You survived getting shrunk!…Uh, which was totally what we expected, by the way. Okay, now you are the size of a microbe. Red blood cells are as big as boats, skin cells are like houses, and a single grain of sand is practically a city! You are the perfect size to meet the mini-creatures that live on and in the human body. We call this community the microbiome.

PRONOUNCER
Microbiome =
mye-croh-BYE-ohm

Your microbiome is made up of the tiny living things that call your body home—both inside and outside. There are about *1 trillion* microbes living on your skin and *100 trillion* in your gut. That's a huge number, kind of hard to picture, so imagine this: If each microbe in and on your body were a dog, they would all cover the entire continent of North America.

Microbiome: the sum total of all the bacteria, fungi, and other microscopic life in a particular environment, like the human body

Like, *totally* cover it. That's a lot of dogs, and that's how many microbes you have.

But these microbes are way smaller than dogs, so instead of covering several countries, they just cover you! To understand exactly how small these microbes are, let's give our imaginations another exercise. This time, picture a grain of salt. Tiny, right? Well, these microbes are 600 times smaller than that one grain of salt. Now look at the ridges on your fingertip. Those microbes are so tiny they could use each ridge as a giant, multilane superhighway! Luckily, we've shrunk down to their size, so it's way easier to see what they're up to.

What's That Smell?

You may be shrunk down right now, but you can still smell stuff—and the human body is covered with some pretty stinky scents. There's that super sock stench. That musty morning breath. The appalling armpit aroma. And of course those fantastically funky farts. Now that you are micro-sized, you can see what's really making these odors—bacteria!

You probably think bacteria are bad because they can cause infections, which is true. But they can also be super helpful, like the ones in our microbiome. There's still a lot scientists are learning about the many good things these bacteria do for us, but one thing they seem to do is keep the bad bacteria out.

Bacteria make pretty good neighbors—they usually pick up after themselves and play their music at a reasonable volume.

And in return, we give them a warm place to live and plenty to eat. The bacteria on the outside of your body love to feast on sweat and the oils made by your skin. The bacteria in your mouth love the leftover bits of food that get stuck in your teeth.

Fact-astic Voyage

There are more cells of bacteria in your body than there are human cells (you know, the ones that you make yourself).

Wait, does that mean stinky breath is sort of like mouth farts?

It's best not to think too hard about it.

When these bacteria eat, they break down foods. They keep the parts they need for energy, and the rest gets released as waste in the form of a gas. It's that gas that gives your body all those intense smells. Foot stink and armpit odor come from bacteria. Your farts are just gas released by your gut bacteria that then escapes out your butt! So technically you didn't cut the cheese—they did!

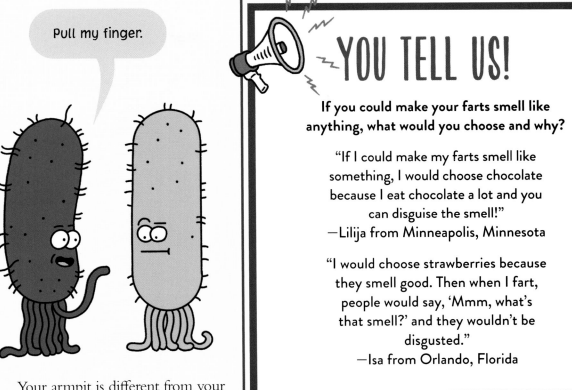

Pull my finger.

YOU TELL US!

If you could make your farts smell like anything, what would you choose and why?

"If I could make my farts smell like something, I would choose chocolate because I eat chocolate a lot and you can disguise the smell!"
—Lilija from Minneapolis, Minnesota

"I would choose strawberries because they smell good. Then when I fart, people would say, 'Mmm, what's that smell?' and they wouldn't be disgusted."
—Isa from Orlando, Florida

Your armpit is different from your foot, so it makes sense that they attract different kinds of bacteria. Different bacteria give off different smells, and that's why your feet and pits each have their own signature stink.

- **Oniony armpit smell** comes from chemicals called thioalcohol. This is made by the bacteria that eat sweat produced in the armpits of teens and adults. Younger kids

MYSTERY PHOTO

Focus your eyes on this mystery photo. Can you guess what it is? Turn to page 127 for the answer.

make a different kind of sweat, which doesn't attract the bacteria that make this onion smell.

- **Eggy fart smell** comes from hydrogen sulfide chemicals. A lot of foods you eat have sulfur in them—like broccoli, chicken, onions, and cheddar cheese. When the bacteria in your gut help break down these foods, they make this distinctive stink.

- **Cheesy foot smell** comes from stuff called methanethiol. This is made when bacteria on your feet eat dead skin cells.

Without the bacteria in your gut, you wouldn't be able to digest many foods. The bacteria living in your intestines help break down all sorts of sugars, fats, proteins, and fibers, and they also help you get the nutrition you need. Sure, they make stinky butt gas in the process, but it seems like a fair trade, right?

Historical Hoax

Joe Rwamirama was a hero in his hometown of Kampala in Uganda, all thanks to his epic farts. They weren't just stinky, they were special—his farts killed mosquitoes. One toot from Joe and the flying pests would drop dead. Eventually, people at a bug spray company got wind of this and asked Joe if they could study him to find out the secret to his fatal farts. Newspapers around the world picked up the story and ran headlines like "Man's farts so deadly he kills mosquitoes in a 6 meter radius." Except none of it was true. The original story came from a news-parody website that makes up stuff to get laughs. And this happened in the year 2019! It goes to show you that we still have trouble sniffing out a good hoax.

SUPER COOL SCIENTIST
DR. HEIDI KONG

Heidi Kong is a dermatologist and researcher at the National Institutes of Health working to better understand the microbiome on our skin. She thinks it's interesting that there are different combinations of bacteria on our skin depending on which part of the body she's studying. For example, the mix of bacteria on your forehead is different from the mix of bacteria on your arm, your mouth, or your toenail. Some patients with skin diseases have different mixes of bacteria, fungi, and viruses on their skin, which is another thing Heidi and her research partners are studying.

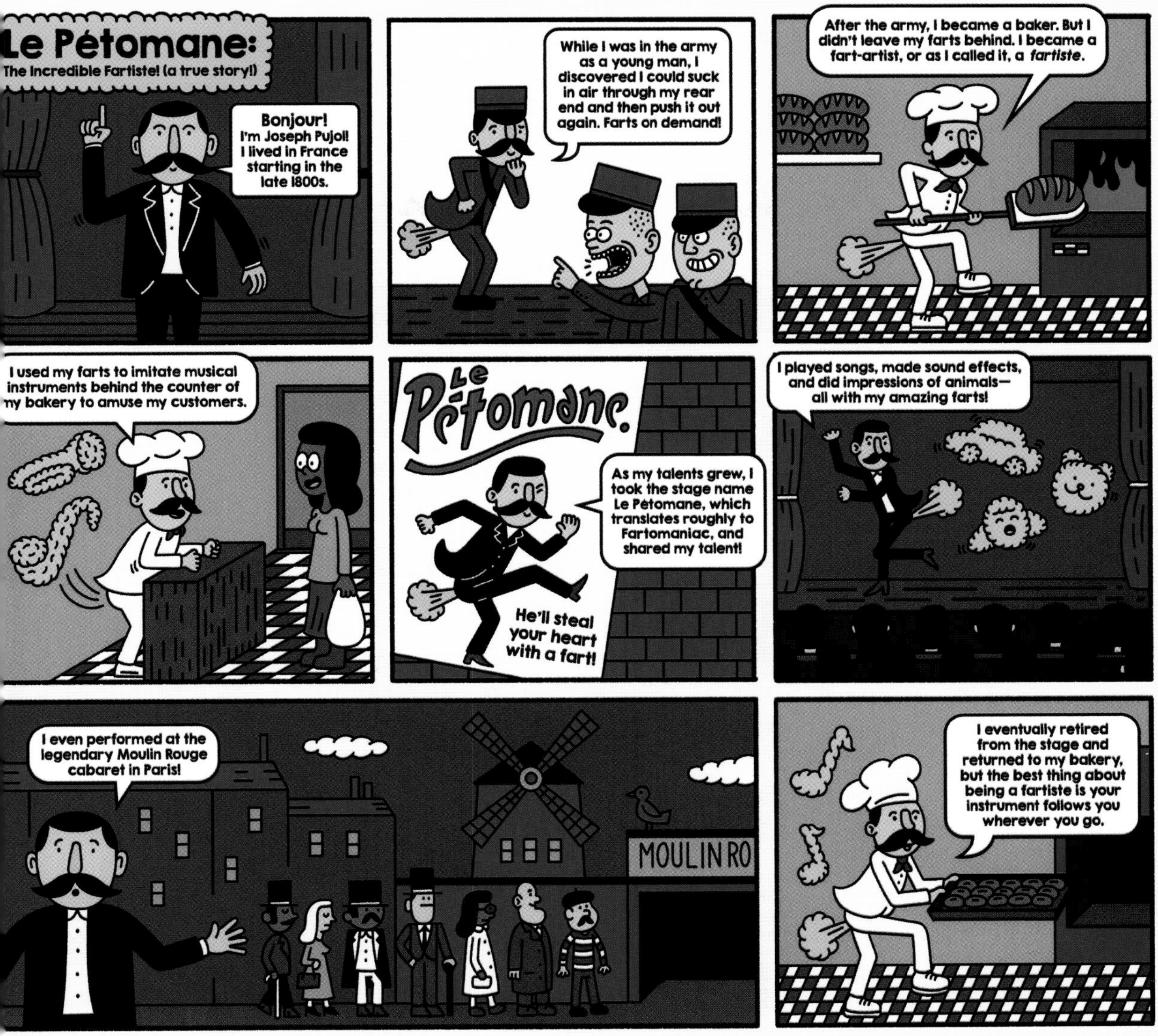
Le Pétomane:
The Incredible Fartiste! (a true story!)
Bonjour! I'm Joseph Pujol! I lived in France starting in the late 1800s.
While I was in the army as a young man, I discovered I could suck in air through my rear end and then push it out again. Farts on demand!
After the army, I became a baker. But I didn't leave my farts behind. I became a fart-artist, or as I called it, a *fartiste*.
I used my farts to imitate musical instruments behind the counter of my bakery to amuse my customers.
Le Pétomane.
As my talents grew, I took the stage name Le Pétomane, which translates roughly to Fartomaniac, and shared my talent!
He'll steal your heart with a fart!
I played songs, made sound effects, and did impressions of animals—all with my amazing farts!
I even performed at the legendary Moulin Rouge cabaret in Paris!
MOULIN RO
I eventually retired from the stage and returned to my bakery, but the best thing about being a fartiste is your instrument follows you wherever you go.

ANSWER!

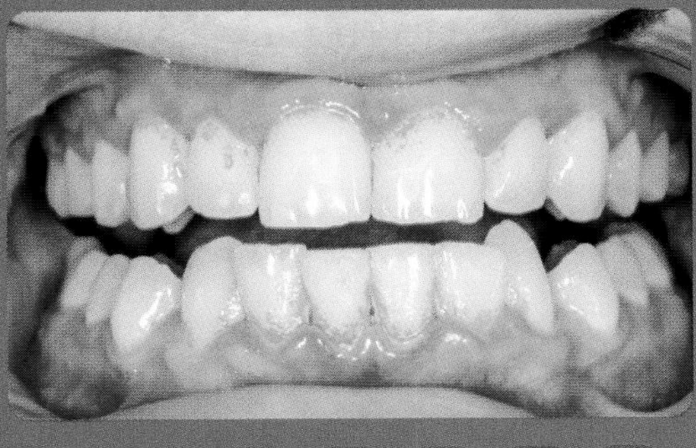

It's plaque! This paste-like substance is a film of bacteria that develops on your teeth and gums. If too much plaque builds up on your teeth, it can harm them and cause them to decay. That's why you need to keep disrupting the bacteria in your mouth by brushing and flossing.

The Tiny Bugs on Your Skin

Let's leave those gassy bacteria behind and meet another major part of your microbiome: teeny-tiny arthropods called mites. Normally you can't see these creatures without a microscope, but when you are shrunk down, you can see that these eight-legged little critters look a lot like their bigger cousins, spiders and ticks. Except mites don't live on webs or in the woods. They live on your face!

Mites like to eat the oil that comes from your skin, so naturally you find them on the oilier parts of your face like near your eyes, nose, and mouth. These Demodex mites spend most of their time hanging out in your pores—those are the tiny holes in your face where hairs, sweat, and oil come out.

Their favorite spot is by your hair follicles (you have tiny hairs all over your body, so this isn't just your head hair we're talking about!). At night, the mites come out on your skin to mate and then head back down into your pores to lay their eggs. Which means there are mite eggs in your face right now. Feeling itchy yet? Mites live only about two weeks, but this keeps a steady supply of them living on you at all times.

Demodex: a group of species of mites that like to live in and around hair follicles. There are species that live on humans and species that live on other mammals.

MOMENT OF EWW

ANOTHER MITE BITES THE DUST

Okay, so maybe you're saying, "Isn't this whole section one big Moment of Eww?" But wait until you hear about dust mites! Dust mites are similar to the ones living on your skin, except instead of eating oils, they love to eat dead skin cells that you and your pets leave lying around the house. And there are lots of them! In one gram of dust, you can find thousands of dust mites. And when people have a dust mite allergy, it's the proteins in the dust mites' poop they're allergic to!

By now, you might have a serious case of the creepy-crawlies, but these mites don't mean any harm. Turns out, you're an awesome place to live! Because mites and humans have been living together for a long time, probably as long as our species has been around, scientists can actually learn about where your ancestors are from by looking at the DNA of your mites!

I like to think of my skin mites as the teeniest of pets. They're helpful, pretty cute, and best of all, they never need a walk!

The Fungus Among Us

Fungus is also a big part of our microbiome, but it hasn't been studied as much as our bacteria, so there's a lot we don't know about it.

But there is one way fungus has really helped humans: penicillin! The bacteria in and on our bodies are very important and helpful, but bacteria can also cause infections. This is when a bunch of harmful bacteria grows out of control in a part of your body. For a long time, humans had no reliable way to stop serious infections from spreading. But in 1928, a scientist accidentally discovered that a kind of fungus called *Penicillium notatum* had a special power: It killed bacteria. Over the next twenty years, many scientists worked to turn this powerful fungus into the medicine now known as penicillin. It was the very first antibiotic, and it saved countless lives. A high five for fungus!

PRONOUNCER
Penicillin =
pen-ih-SILL-in

PLURAL PLURALS

If you see more than one fungus, use the plural "fungi"... or don't, because you can also make it plural by saying "funguses." Both are correct. And to make it more confusing—I mean, fun—you can pronounce the first one two ways: FUN-jye or FUN-guy. Basically, you're never wrong!

CLOUDY WITH A CHANCE OF MICROBES

Our microbiome isn't just on us...It's also floating around us. Scientists have discovered that there is an invisible cloud of bacteria, fungus, and dead skin cells hovering outside each and every one of us. When we move, we send more of these microbes flying off our skin, and since they are so light, they just float in the air! This cloud gets on the furniture, walls, and people near us, even if we never touch them. So if someone ever tells you you're living with your head in the clouds, tell them, "Yeah, in fact my whole body is in a cloud!"

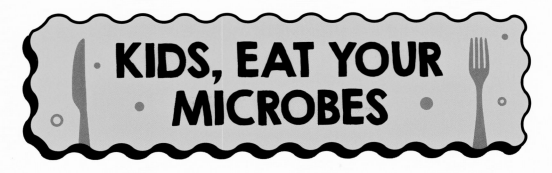

KIDS, EAT YOUR MICROBES

Hungry for a helping of bacteria? How about some fennel and fungus soup? Or maybe a little bite of mite? These meals might sound totally *ick*, but microbes are actually a big part of the foods we eat. Don't worry, most of them are harmless. Some even help make our favorite dishes extra delicious. So grab a plate and your appetite! We're going to chew our way through the world of micro bites! CHOMP!

A PINCH OF SUGAR, A DASH OF BACTERIA

Bacteria and fungi are two of the world's greatest chefs, but they rarely get the love and respect they deserve. Both of them help create foods we know and love. For example: Do you like bread? That's chef fungus! Do you like chocolate? That's chef bacteria. Do you like pickles? That's also chef bacteria! Do you like chocolate-*covered* pickles? That's just gross.

What?! Chocolate-covered pickles are the absolute best! Don't yuck my yum.

These two micro master chefs work by transforming ingredients into tasty foods using a process called fermentation.

Let's take dill pickles as an example. They start as cucumbers that are naturally covered in bacteria, including one called lactobacillus. The cucumbers are plunked into a mixture of salt and water

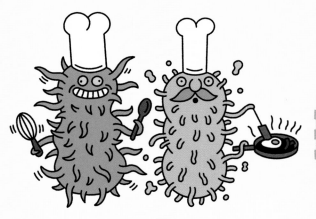

Fermentation: when bacteria or fungi eat up sugars and carbohydrates in food and then turn those things into something new, like alcohol, acids, or gas

131

called brine. That's when chef lactobacillus goes to work.

The lactobacillus bacteria start by eating up the natural sugars in the cucumbers. Then they release stuff called lactic acid. This stuff is amazing. First, it stops any bad bacteria from growing, so your pickles can last years in a sealed jar without spoiling! Second, lactic acid adds a tartness and tang to the pickles, giving them that sour taste. And get this: Some lactobacillus bacteria even make B vitamins when they are fermenting those pickles, adding a healthy boost.

When it comes to fermentation, pickles are just the tip of the snack-berg. Fermentation is also used to make sauerkraut, sour cream, kimchi, salami, miso, yogurt, tempeh, Tabasco sauce, dosa, wine, beer, vinegar, and so much more. Even the cocoa beans that eventually become chocolate must first be fermented to make them less bitter and to bring out their delicious flavor. It's no wonder there are so many fermentation fanatics out there.

Yeast Mode!

Baking bread is like a magic trick. You take plain water and some dry powders, mix them, stick them in an oven, and TA-DA! Out comes fluffy, chewy, scrumptious bread! It's quite a transformation, and the magician behind this presto change-o is a microbe called yeast.

Yeast is a fungus, and like the lactobacillus bacteria, it also kicks off fermentation. In this case, the yeast chows down on the dry powders, like flour or maybe some sugar. But instead of releasing lactic acid, yeast makes alcohol and a gas called carbon dioxide. The alcohol evaporates during baking, and the carbon dioxide gas fills up little pockets in the dough. Over time, those holes start to expand as more gas gets in there—sort of like blowing up a lot of little balloons inside the dough. This makes the dough grow and rise and gives bread that extra-fluffy texture after

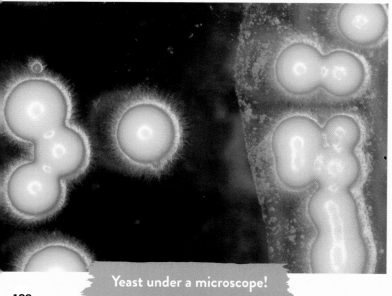

Yeast under a microscope!

baking. Not all baked goods use yeast to rise, but most breads do. So next time you bake bread, picture a lot of tiny yeast magicians helping make it rise. *Alacazam-wich!*

And for my next trick, I will make this bread disappear!

Who Invented Bread?

On a dark and stormy night, a brilliant scientist named Dr. Belinda Bakerstone was tinkering in her lunch lab when lightning struck her water jugs, flour bags, and yeast, and it formed a monstrous new creation: bread! Okay, just kidding. No one really knows who invented bread. We *do* know that bread needs only three ingredients: flour, water, and yeast. Humans have been grinding up grains to make flour for at least 22,000 years! Scientists discovered a super old stone from back then that was used to grind barley.

It's likely those ancient people added water to that barley flour and ate it like a porridge. If they accidentally let that porridge sit out too long, natural yeasts in the air could land on it and start fermentation. The porridge would then get a little sour. If people decided they wanted to reheat that old sour porridge, they would put it near a fire—and suddenly it would get harder and fluffier, becoming…bread!

Scientists think the first bread was probably made like this: totally by accident! And it probably happened lots of times in different parts of the world before humans began really perfecting the art of baking. From what we know, the earliest breads were flat, like pita bread, tortillas, or naan. Over time, different cultures developed different ways to bake, and now you can find unique breads around the world! All because some people

HANDS OFF MY SLICED BREAD!

Have you ever heard someone say, "That's the best thing since sliced bread"? Well, sliced bread wasn't always so popular. In the 1920s, a guy named Otto Rohwedder invented a machine to slice a whole loaf of bread at a time. Many bakers thought it was a terrible idea. "The bread will go stale, and it will fall apart in the bag!" they said. But over time, it went from a strange new invention to the best thing since…well, just the best thing. During World War II, the American government banned sliced bread so the supplies used to make it could help with the war instead. People were so upset they wrote letters to complain! The ban lasted only two months.

left their porridge out too long. Sometimes it pays to be a slow eater.

Microbes + Milk = True Love

Microbes love milk. Sometimes they spoil the milk, making it sour, lumpy, moldy, and gross. But in the right situation, they can turn milk into fantastic foods, like yogurt, cheese, sour cream, or buttermilk. Yogurt, for instance, is made by heating up milk to kill off all the bad bacteria, then adding in good bacteria to ferment it. Not only do these good bacteria give yogurt a refreshing, tangy taste, they also help with your digestion. You see, some of these good bacteria stay alive long after you eat them. They will travel with the yogurt into your intestines, where they may stick around to help you break down food. So when you eat yogurt, it's kind of like you're inviting tiny new friends to live inside you. The more the merrier, right?

Cheese also comes from fermented milk, but the way you make it is a little more complicated.

1. **Cheesemakers add bacteria and other ingredients to milk so it gets nice and clumpy.**
2. **The extra liquid is drained and salt is added to those clumps, which are called curds.**
3. **Next, the clumps are squished to get rid of extra moisture so they become nice and firm.**
4. **Finally, the cheese is put in a cool, dark room to age. Some kinds of cheese are aged for only a few weeks, while others are aged for years.**

Bacteria are active the whole time, eating up sugars from the milk and releasing acids that give the cheese that cheesy flavor. But bacteria aren't the only microbes at work. Let us introduce you to some tiny champions of cheese.

MOMENT OF EWW

THE YOGURT CURE

Legend has it that yogurt once saved a king from an embarrassing potty problem. In 1542, the king of France was in trouble. He had a bad case of diarrhea that just wouldn't go away. French doctors tried everything but alas, the king just kept pooping! It was a hopeless, stinky situation. Then Suleiman the Magnificent, who was the ruler of the Ottoman Empire, sent one of his doctors to help. That doctor gave the king a food that was new to France, but very popular in the Ottoman Empire: yogurt. The king ate it up, and soon his toilet troubles were over! The king and his royal rump were very grateful.

MEET THE CHEESE MITES!

Hi, I'm Chelsea. And this is Chad. We're cheese mites.

We poop in your cheese!

Wait, wait, whoa...don't freak out. It's safe to eat! In fact, some cheesemakers actually add extra mites to their cheeses. You see, when we eat our way through a cheese, we excrete digestive juices that add extra flavor.

We're delicious!

Yeah. We're even an important part of fancy cheeses like Milbenkäse and Mimolette.

Plus, who cares if mites poop in your cheese? There are mites pooping everywhere—even on your face right now! Bye!

Historical Hoax

In the late 1950s, there was an Italian company selling bags of grated Parmesan cheese. Thousands of Italians ate it up. But it turns out it wasn't actually Parmesan...or cheese. Food inspectors found out that the company was really selling old, ground-up garbage mixed with a glue-like material used to make buttons, umbrella handles, and toys. Officials seized 6,000 pounds of this fake cheese. Apparently, the people who ate this stuff were fine with it, because the company got very few complaints.

Why Does Food Go Bad?

Deep in your fridge, hidden behind the butter, it waits, it lurks, it's coming to make you spew! It's…MOLDY BOLOGNA!

Not all microbes are helpful. Many can spoil your food and make you sick if you eat them. We keep food cold to stop these putrid pests from spreading, but they don't give up easily. With enough time, microbes can spoil pretty much any meal, so it's best to be on the lookout for their telltale signs.

THE "IS THIS OKAY TO EAT?" CHECKLIST!

☑ Is it lettuce or deli meat with a weird, slimy film?

☑ Is it a fruit or vegetable that is suddenly mushy and brown?

☑ Is it bread or cheese with green fuzz?

☑ Is it milk or soup that smells like pure death?

☑ Is it a long-forgotten open jar of spaghetti sauce that is suddenly self-aware and plotting to take over the kitchen the minute you turn your back?

If you checked any of the boxes…DO NOT EAT THAT FOOD! Send it straight to trash-town.

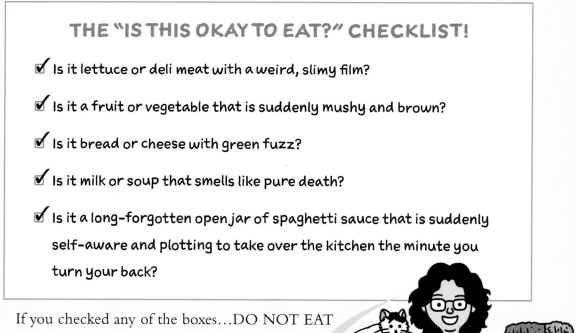

Remember! Fuzzy cats = good!
Fuzzy bread = bad!

Since you are shrunk down to microbe size, you can see something other people can't: that even fresh foods often have some amount of bad bacteria and fungus on them. Usually these amounts are so low that we don't get sick if we eat them. The trouble starts when those bacteria or fungi have time to grow and spread across the food.

A bunch of bad bacteria might multiply in some old soup, feeding on the noodles and releasing toxins into the broth. The more time that passes, the more toxins fill the soup. Even though it may look totally normal, it might smell a little funky. Eventually, there will be enough bad stuff in there that if you ate the soup, you'd be very sick, super quick.

Mold is a type of fungus that can grow on food too. It spreads by launching lots of little spores into the air, like someone throwing a bunch of paper airplanes into the wind. These spores are sort of like mold seeds. Some of them could land on your pizza without you noticing. Given enough time, the spores will grow and spread more mold across the pizza, adding a new topping to it: fuzzy green lumps. Eating mold can seriously upset your stomach. Some types even make dangerous toxins. And don't try cutting off the moldy parts and eating the rest! Mold grows long, thin roots deep into food. So there may be more mold than meets the eye.

One of the best weapons against these microbes is your refrigerator. When microbes are cold, they spread much more slowly, giving you more time to eat those leftovers before they spoil. Canned foods last longer too, because cans are sealed with no air inside, and many types of bacteria need air to spread. But—spoiler alert—in the end, microbes will probably ruin your food. So be alert, look for mold, and when it comes to leftovers, always sniff before you snack.

Fridge, you saved my lasagna! You're my hero!

MICROBES GET EXTREME

While you're still shrunk down to microbe size, why not take a micro vacation? A beach day, maybe? A national park? Perhaps a trip to the mountains? Well, be sure to buckle up, stow your luggage, and get your passport stamped when you enter the outdoor worlds of the micro-verse. You can probably name at least a hundred different plants and animals that live out there, but when the Zoom Ray is involved, that number grows and grows as you shrink and shrink. In fact, there's an incredibly vast number of microbes covering the ENTIRE planet. It *IS* a small world after all.

THE TINY ORGANISMS UNDER THE SEA

Grab some goggles, snatch a snorkel, and get ready to do the bacteria backstroke, because we're going swimming in the micro ocean. When you're regular size, you might find dolphins, octopuses, whales, and snails! But when you're the size of a cell, there's a whole other group of organisms to discover. From the ocean floor to the cresting waves, microbes cover every inch of the salty seas. In fact, if you scooped up a liter of seawater, you would be holding more than 1 billion microbes!

Let's swim deep down to the bottom of the ocean. You see those bubbling hot cracks in the ground? Those are called hydrothermal vents. They pop up in places where the earth's tectonic plates are moving apart and exposing mineral-filled magma underneath. Hydrothermal vents can get very hot—up to 750 degrees Fahrenheit as they shoot streams of scorching water up from the ocean floor. Since they are so deep down, there's very little oxygen and no sun around them. What could possibly live in this extreme environment? It's our tiny pals: bacteria!

> **Tectonic plates:** Colossal slabs of rock that make up the outer layers of earth's crust. They are in constant slow motion, scraping up against each other and pulling apart.

138

But these aren't your average single-celled organisms, not by a long shot. These bacteria are super tough and apparently…super tasty? Since the discovery of hydrothermal vents, more than 800 new species of animals have been found feeding on these microbes, like ghostly

Fact-astic Voyage

Many scientists think that life on earth began about 3.7 billion years ago at hydrothermal vent sites. What started as tiny microbes evolved into not-quite-as-tiny microbes. Some ate the others as food and kept evolving, until eventually whole new species of animals emerged. It took more than 3 billion years to get from microbes to animals walking on two legs.

yeti crabs, tube worms the size of an average human, and clams as big as your head!

Plankton: The Ocean's Chillest Microbes

Let's leave those toasty vents and swim closer to the surface of the ocean. Here you'll find plankton. Technically, to be classified as plankton, you simply have to live in a body of water, like the ocean, and get from place to place by drifting along aimlessly with the currents. So that means jellyfish rolling in the waves are plankton. Same goes for baby shrimp, although when they grow up and move on their own, they are no longer planktonic. Mostly plankton, though, are teeny-tiny organisms not even visible to the naked eye. Examples include algae, bacteria, crab-like crustaceans, and tiny snails. Are they lazy? Are they super chill? Did they just never learn to swim? Who knows—but one thing is clear: Without plankton, ocean life would be in trouble.

Just like microbes at hydrothermal vents, plankton are often at the bottom of the food chain, which means they help feed lots of creatures. Think of them as your lunch's lunch's lunch's lunch. Plankton can be divided into two types: plantlike phytoplankton and zooplankton, which are animal-like organisms.

Creatures big and small feast on plankton. Fish love them. So do crabs and penguins. Even whales eat plankton. In fact, pretty much the *only thing* blue whales eat is krill, a shrimplike zooplankton (see page 28 for more on whales and krill). Given that blue whales are the largest animal on earth and can be the size of three buses, they need A LOT of food. A typical blue whale can chow down around 8,000 pounds of krill a day—that's about 40 million tiny animals.

PRONOUNCER
Zooplankton =
ZOH-oh-PLANK-tun

Fact-astic Voyage

Some single-celled phytoplankton called dinoflagellates glow neon blue when they get sloshed around. The color change is called bioluminescence, and it's a chemical reaction inside the cell. It's the same process that makes fireflies light up. When enough dinoflagellates are in the same place, the water can start glowing!

DO WE BREATHE TREES OR SEAS?

The air you're breathing right now likely came in large part from phytoplankton like algae. Scientists estimate that 80 percent of the earth's oxygen comes from these single-celled organisms! Like other plants, they convert carbon dioxide into sugars and oxygen, a process known as photosynthesis.

A RIVER RUNS RED

If you take a trip to southwestern Spain, you may find yourself staring at a river of red water: the Río Tinto. The red color comes from large amounts of iron and sulfur in the ground, which make it practically toxic for life. But that doesn't stop our micro pals like bacteria, algae, and fungi. Scientists have found all these and more thriving in these iron-infested waters. The bacteria have been nicknamed "rock eaters" because their diet is made up of bits of iron in rocks. It turns out the surface of Mars is similar to the Río Tinto, so astrobiologists from NASA are busy studying this microverse for clues about what life on another planet might be like.

CAVES, MORE THAN JUST A HEADQUARTERS FOR SUPERHEROES

Dry off and grab a flashlight, because we're heading to our next micro destination: caves! They're the home of bats and Batman. They're cold and creepy but also full of cool rock formations. Cave explorers can travel for days into these pits of darkness, reaching secret rooms far below the surface of the earth. What do they find there—cool rocks and crystals? Yep. Buried treasure? Only if they are super lucky. Bacteria? Yeah, lots and lots of them, and some have superpowers.

In 2002, researchers exploring Kentucky's Mammoth Cave grabbed some bacteria samples to study at their lab. One of the bacteria they brought back produces a substance that can make it harder to grow new blood vessels. This is a big deal, because cancer tumors use these sorts of blood vessels to grow. So the bacteria could one day help create treatments that fight tumors and save lives!

SUPER COOL SCIENTIST

DR. HAZEL BARTON

Dr. Hazel Barton is a microbiologist and climbs into caves all over the world looking for new life. Like any biologist studying tiny organisms, she uses powerful microscopes, but her equipment also includes helmets, lanterns, ropes, scuba gear, ice boots, and the occasional kayak. Getting the perfect specimen a thousand feet belowground can sometimes put her in very small spaces. She says the smallest space she can squeeze through is about six and three-quarter inches. According to her, the trick is to let all the air out of your lungs, so they get much smaller, and push through to the other side.

MOMENT OF EWW

THIS CAVE NEEDS A TISSUE

It's not because of a cold, but the result is the same. Some cave walls ooze with snottites, a goo that, over time, looks like droplets of snot. This globular gunk is filled with microbes, and the mucus-like film is so acidic that it could burn through your clothing and do real damage to your skin. Stay away from this cave snot!

Some of us want to stay away from *all* snot.

A Not-So-Friendly Fungus

Bats might be the most famous cave creature. Many species depend on caves for shelter, especially during the cold winter months, when they hibernate. But in North America, bats are under attack, and the culprit comes from the micro-verse.

White-nose syndrome comes from a fungus that loves to grow in cool, moist places like caves. The fungus attaches itself to bats' noses and wings and causes our mammalian friends to wake up too early from hibernation, when it's still cold and there's not enough food outside. Unfortunately, this causes some bats to die, and white-nose syndrome has become a huge threat to species like the little brown bat.

Even if you aren't a fan of bats themselves, you are probably a fan of what they do for the planet. Some spread seeds for trees, while others pollinate plants that grow many of our favorite fruits. Bats also help keep the insect population down by eating tons of bugs like mosquitoes. And don't forget about bat poop—also called guano. Bat poop is such a prized fertilizer, countries have gone to war to get it! So scientists are working hard to find a way to stop this bat-killing fungus and save our flying friends.

FROM TINY SPORES TO WHOA THAT'S A GIANT MUSHROOM

Now let's get some fresh air and take a hike in a forest of fungus! There could be more than 5 million different species of fungi! So there's plenty to see in this microworld. You probably know the bigger stuff, like mushrooms, but that's only the beginning. You can basically find fungi in every corner of the globe. So let's meet the fungus among us!

Trees and bushes are happy taking center stage in the woods, but true forest-fans know the real stars are the fungi. If it weren't for them, the forest wouldn't be able to keep growing. Let's say a tree dies. Fungi will grow on the dead wood and break down all the leaves and branches and roots until eventually the tree is gone. What's more, as fungi make a meal of this tree, it leaves behind nutrients that go back into the soil and help spawn new plants. To completely break down something as big as a tree takes decades, but without fungi it wouldn't happen at all. In order to keep this cycle going, mushrooms need to keep making more of themselves. Just like the mold you read about earlier, mushrooms and other fungi release seedlike spores to keep growing.

For an extreme case of spore-mageddon, let's pay a visit to one very special mushroom, a honey mushroom. Like all mushrooms, it started out as a spore. It found a cool, damp home in an Oregon forest. Except it wasn't called Oregon when this spore started to build its home more

Spore: A small particle made by many fungi to reproduce. Unlike plant seeds, a spore does not need to be fertilized or pollinated and can produce a copy of the fungus on its own.

than 2,000 years ago. The land was overgrown, and the northwestern climate was just right for this spore to grow into a mushroom and send out new spores. When a honey mushroom meets another with the same genetic makeup, they can fuse together. Over

MYSTERY PHOTO

Focus your eyes on this mystery photo. Can you guess what it is? Turn to page 145 for the answer.

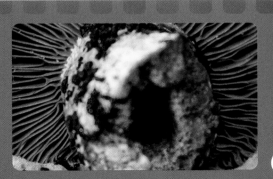

Fact-astic Voyage

Spores nestled in the poo of farm animals are the fastest-traveling organisms in the world. Using a high-speed camera, researchers discovered that the spores of the dung cannon, or hat thrower, fungus can accelerate at one hundred times the speed of sound. They power this supersonic trip by building up water underneath the spore tip, which looks a bit like a floating black eyeball. The water bursts the spore off its poo host and onto a nearby patch of grass. The spores need to be launched with such speed in order to land somewhere other than where they came from.

the years, this particular honey mushroom has become a master at this fusion trick, and it has grown into an empire. Covering 3.4 miles, it's the largest living organism on land. It's safe to say this mushroom takes up much room.

MATCH THE MUSHROOM
Use the names below to identify each mushroom.

Wrinkled peach • Brain mushroom • Bioluminescent • Bearded tooth mushroom

A. Bearded tooth mushroom; B. Wrinkled peach; C. Bioluminescent; D. Brain mushroom

Farming Friends and Frenemies

Lots of creatures eat fungus, and some even farm it—like the leaf-cutter ants living in the South American rain forest! These ants chop up certain leaves and bring them home. Similar to farmers who harvest hay to feed cows, these ants use the leaves to feed fungus! The fungus grows and grows as it gobbles up those bits of leaves. Eventually, baby leaf-cutter ants come to eat the fungus. It's a symbiotic relationship. That's when two organisms—like the ant and the fungus—work together. In this case, the fungus gets to eat leaves and the ants get to eat some of the fungus. Win-win!

But not all relationships with fungi are good for everyone involved, like the creepy case of the zombie spores! *Ophiocordyceps unilateralis* is a fungus that lives in tropical forests. It likes to attach itself to ants and slowly take over their bodies. At first, the ant may seem fine, but as the fungus spreads inside the insect, the ant will start acting weird—leaving the other ants and attaching to the underside of a tall leaf. Once there, it dies, and eventually a long stalk of fungus will grow from its head and launch more spores. Those spores will float off, looking for more ants to infect.

PRONOUNCER
Symbiotic =
sim-bee-AH-tick

The micro-verse is equal parts creepy, cool, weird, and wonderful. There's a lot to see here, but one thing it doesn't have: chairs. Don't you miss chairs? So big. So comfy. So sittable. Hmm…maybe it's time we un-shrink so we can sit down for a bit. Let's fire up the Zoom Ray…set it to grow mode and…*ZHWUH-VOOOOOOM!* Ahhhh, there we go—back to normal size. But never forget what we learned today: There are tiny creatures everywhere—in nature, in your food, and pooping right now on your face!

ANSWER!

They're the gills on the underside of an indigo milk mushroom! If you were to walk by this mushroom and look down, it would be a pretty unremarkable white cap you might see in any patch of grass. But turn it over, and marvel at this blue beauty! And if you cut it, it emits a blue substance. A true blue blood!

145

MEGA MATCHUP
TARDIGRADE VS. SLIME MOLD

In today's tiny takedown, we have two outrageous organisms ready to go head to... slime? In one corner, we have the hearty, indestructible, and adorable tardigrade! And in the other corner, we have the astonishing, mysterious, and blob-erific slime mold! Which organism will come out on top?

TEAM TARDIGRADE

- This micro animal, also known as a water bear, can be found living pretty much everywhere: deep in the ocean, on the tops of mountains, and even in backyards and parking lots. And people are finding new species of tardigrades all the time!

- Tardigrades can survive the most extreme environments: They can be in places as cold as minus 328 degrees Fahrenheit or as hot as 304 degrees, and they can go thirty years without water.

- You can send tardigrades to explore space without any protective gear! They can survive in that cold, dark vacuum for ten days—meaning they can withstand extremely high levels of radiation and very, very high pressure. Good thing, since it would be hard to make a space suit that small.

- These adorable little creatures are ancient. They pre-date the dinosaurs and have been around for at least 500 million years!

- One of the keys to their survival is they can go into a state where they shed their water, curl up into a tiny ball, and basically shut down their entire body. It's still running but very, very slowly. They can be in this state for decades and reanimate when they come into contact with water.

146

TEAM SLIME MOLD

- Don't let the name fool you—slime mold isn't a mold at all. Slime molds are single-celled living things that can be microscopic. They can find one another and merge into a much larger single-celled life-form that can be bigger than a person.

- And these slime molds can move! Just very slowly. They travel by changing their shape, crawling, and moving tendrils. They can even break apart and merge back together again.

- They can sense chemical signals and then move either away from them or toward them, depending on what the signals are. This ability is called chemotaxis. Slime molds also leave chemical trails behind them, which in a way gives them "memories" of where they've been.

  - Even though these slime molds are made up of a single cell and have no brain, they can do some pretty amazing things! They can navigate mazes and escape from traps. They've even been used in microchips to drive robots and toy boats.

  - There are lots of intriguing questions to answer about slime molds, but one reason scientists are studying them is that they hope to build robots modeled on them. These robots one day might even be able to travel inside our bodies and help perform surgery!

Which tiny organism is cooler: Tardigrade or Slime Mold?

YOUR VERDICT

MOMENTS OF UM

WHY ARE THERE HOLES IN SWISS CHEESE?

Cheese is full of microbes. They live in the cheese, happily munching on its sugar and fat. Just like us, they pass gas. In this case, it's a gas called carbon dioxide. Since they are tooting inside the cheese, this gas is trapped. It blows up the insides of the cheese, like little balloons. When those balloons pop, they leave behind holes. Cheese experts call these holes "eyes."

WHY DOES YOUR BREATH SMELL WORSE IN THE MORNING?

Saliva is very useful for keeping your mouth smelling good. It helps sweep away particles of food and bacteria. When we sleep, we don't produce as much saliva as we do as when we're awake, and this makes our breath smell worse in the morning. If you sleep with your mouth open, it can make your mouth drier and the morning breath even stinkier.

WHAT'S THE DIFFERENCE BETWEEN FARTS AND BURPS?

Both are caused by gas, but they have different sources. Your burps come from gas in your stomach or esophagus. Most of this is air you've swallowed while eating and drinking, or carbon dioxide from fizzy drinks (the fizziness is made with bubbles of gas!). Farts, as we mentioned earlier, are produced by the bacteria in your intestines.

IS IT TRUE THAT MAKEUP HAS BUGS IN IT?

If you look at the back of makeup that has a red tint, you may see the word "carmine" on the ingredient list. Carmine is a pigment that comes from the female cochineal insect, and it's been used for thousands of years because it has a deep red color. These insects contain about 20 percent carminic acid, which is the source of the color. To get the acid, you need to crush up the dead bugs, soak them in an alcohol solution, and filter out the insect parts, and then you're left with this carmine pigment. It's more or less bug juice.

ARE VIRUSES ALIVE?

This is something that scientists don't agree about. Some argue, "Yes, they are living things!" and others say, "No way!" This is because viruses do not eat and cannot live outside the cells of their host. But they can make copies of themselves and have a big impact on their surroundings. So viruses exist in a gray area between living and nonliving.

WHAT IS DUST MADE OF?

Those dust bunnies under your bed aren't alive, but we're responsible for making them. A lot of the dust in your house is made up of dirt and sand brought in from outside, and it also includes hair and dead skin cells (from you and your pets), fibers from clothes and furniture, dust mites (and their poop), and pollution from cars, construction, and factories. The mix of particles in the dust in your home is unique and depends on who lives there, where your home is located, and what's around it. It's almost like a fingerprint!

DO ANIMALS HAVE MICROBIOMES?

Yes! Humans aren't the only ones with microbiomes. Mammals, reptiles, fish, plants, and insects all have their own communities of microbes living in and on them. In fact, studies have found that the microbiomes of humans and their pet dogs become similar to each other over time.

WHAT IS THE TINIEST THING A MICROSCOPE CAN SEE?

Some of the earliest microscopes used in the 1600s helped scientists discover bacteria, nematodes, and other tiny living things. These devices use light and lenses to zoom way in on things. But now there are microscopes that use electrons instead of light to zoom way, way in. These electron microscopes are able to show us individual atoms—the eensy-teensy, super tiny building blocks that make up everything.

THE END

Maybe something about how amazing life is. After all, we learned about animals that can change color and shape and plants that communicate with one another. We saw that life can exist pretty much anywhere, and we learned that there's way more going on around us (and on us!) than meets the eye. And that's barely scratching the surface! There's so much more to discover.

So when you close this book, keep the questions coming. Look around you. Dig in the dirt. Watch the sky. Spy on plants. Study your pets. Keep tabs on birds. Make friends with fish. When you stumble across something you don't understand, go hunt for answers—you'll be glad you did. This world is full of wonder and adventure, and all you need to explore it is an open mind. A Zoom Ray helps too, but that's totally unnecessary.

ACKNOWLEDGMENTS

There are so many people we want to thank. Without them, there would be no book!

High fives to Menaka Wilhelm, who is funny, brilliant, and makes our podcast better; Elyssa Dudley, who generously lent us her brainpower and talents; Tracy Mumford, who provided sage counsel and encouragement; and Lauren Dee, who made this all happen.

A big, blue whale–sized THANK-YOU to the people who helped us along the way: Mike Reszler, Lily Kim, Kristina Lopez, Rosie DuPont, Brandon Santos, Lindsey Davis, Tsering Yangcheng, Phyllis Fletcher, Emily Kittleson, Kris Cramer, Ruby Guthrie, Emily Allen, Ned Leebrick-Stryker, Jon Lambert, Corey Schreppel, Veronica Rodriguez, Eric Romani, John Miller, Eric Ringham, Sam Choo, Steve Griffith, and Rachel Dennis.

Thank you to the people who gave this project nudges in the right direction early on: Collin Campbell, Peter Clowney, Hans Buetow, Steve Nelson, Jon Gordon, and Kristen Muller.

We ADORE our editor, Samantha Gentry. She taught us everything about how a book is made and made this one so much better. Special thanks to Albert Lee for being the most enthusiastic cheerleader ever! Many thanks to Laurie Calkhoven, and all those at Little, Brown Books for Young Readers: Megan Tingley, Jackie Engel, Lisa Yoskowitz, David Caplan, Karina Granda, Jen Graham, Andy Ball, Siena Koncsol, Stefanie Hoffman, Natali Cavanagh, Savannah Kennelly, Victoria Stapleton, Michelle Campbell, and Christie Michel. A million thanks to Ryan Katz, Tarri Ryan, and Barbara Natterson-Horowitz for making sure we got it right. Sincere appreciation to Serge Seidlitz and Neil Swaab for bringing this book to life.

We want to thank all the super cool scientists who have been so generous with their time and expertise. You've taught us so much, and it's an honor to share your work with the wider world. And so much love to all the kids who shared their questions and ideas with us. Your creativity and curiosity power everything we do.

And finally, epic thanks and highest of fives to our families for their support. To Andy and Lulu for making home the very best place to be, and to Carolyn, Stuart, Mickey, Delia, Dobie, Dickie, and Leenie for being the loudest, most supportive cheering section. To Vikki and Coco for all the laughs, and for always being available when the podcast needs an extra voice (shhhhh-hhhh!), and to Jody and Skip for hiding their skepticism and never saying no. To Kathy for always believing in this project. To Vicken, Bo, Brendan, and Collin for your support and for sharing our love of big questions. And to Penelope A. Poodle for all the cuddles.

SELECTED BIBLIOGRAPHY

PART 1: ANIMALS

"Caribou." U.S. Fish & Wildlife Service. December 6, 2016. https://www.fws.gov/refuge/arctic/caribou.html

Curtin, Ciara. "Fact or Fiction?: Urinating on a Jellyfish Sting Is an Effective Treatment." *Scientific American*, January 4, 2007. https://www.scientificamerican.com/article/fact-or-fiction-urinating/

Farago, Tomas and Sarah Ellis. "Barks, growls, meows and purrs: Translating cats and dogs." Interview by *Brains On!*, American Public Media, December 8, 2015. https://www.brainson.org/shows/2015/12/08/barks-growls-meows-and-purrs-translating-cats-and-dogs

Fash, Nick. "How do animals breathe underwater?" Interview by *Brains On!*, American Public Media, March 24, 2015. https://www.brainson.org/shows/2015/03/24/how-do-animals-breathe-underwater

Fox-Skelly, Jasmin. "The smelliest animals in the world." *BBC*, June 10, 2015. http://www.bbc.com/earth/story/20150610-why-its-good-to-smell-bad

Gardiner, David. "Healing skin and regrowing limbs: The science of regeneration." Interview by *Brains On!*, American Public Media, October 29, 2015. https://www.brainson.org/shows/2015/10/29/healing-skin-and-regrowing-limbs-the-science-of-regeneration

Gonzalez-Bellido, Paloma and Kate Feller and Sergio Rossoni. "Do insects see the world in slow motion? Looking through animal eyes." Interview by *Brains On!*, American Public Media, July 9, 2019. https://www.brainson.org/shows/2019/07/09/do-insects-see-the-world-in-slow-motion-looking-through-animal-eyes

Hanlon, Roger. "Cuttlefish: Ultimate shapeshifters!" Interview by *Brains On!*, American Public Media, June 9, 2015. https://www.brainson.org/shows/2015/06/09/cuttlefish-ultimate-shapeshifters

Heenehan, Heather. "How do whales communicate?" Interview by *Brains On!*, American Public Media, November 8, 2016. https://www.brainson.org/shows/2016/11/08/how-do-whales-communicate

Helm, Rebecca. "How and why do jellyfish sting?" Interview by *Brains On!*, American Public Media, April 9, 2015. https://www.brainson.org/shows/2015/04/09/how-and-why-do-jellyfish-sting

Herzing, Denise, Michael Zasloff and Alicia Bitondo. "Dolphins vs. Octopuses: Showdown in the sea!" Interview by *Brains On!*, American Public Media, January 30, 2018. https://www.brainson.org/shows/2018/01/30/dolphins-vs-octopuses-showdown-in-the-sea

Horowitz, Alexandra. "Do dogs know they're dogs?" Interview by *Brains On!*, American Public Media, November 6, 2018. https://www.brainson.org/shows/2018/11/06/do-dogs-know-theyre-dogs

Laidre, Kristin. "Narwhals: Unicorns of the sea?" Interview by *Brains On!*, American Public Media, October 10, 2017. https://www.brainson.org/shows/2017/10/10/narwhals-unicorns-of-the-sea

Lisberg, Anneke. "Dogs: What's the secret of their sense of smell?" Interview by *Brains On!*, American Public Media, November 16, 2015. https://www.brainson.org/shows/2014/11/16/dogs-whats-the-secret-of-their-sense-of-smell

McComb, Karen. "Cats: Glowing eyes, puffy tails and secret purrs." Interview by *Brains On!*, American Public Media, July 27, 2015. https://www.brainson.org/shows/2015/07/27/cats-glowing-eyes-puffy-tails-secret-purrs

Oberhauser, Karen. "How do monarch butterflies travel so far?" Interview by *Brains On!*, American Public Media, October 9, 2014. https://www.brainson.org/shows/2014/10/09/how-do-monarch-butterflies-travel-so-far

Rayor, Linda. "Walking on walls: How ants and spiders do it." Interview by *Brains On!*, American Public Media, March 5, 2019. https://www.brainson.org/shows/2019/03/05/walking-on-walls-how-ants-and-spiders-do-it

Unguez, Graciela. "The nerve! Electricity in our bodies." Interview by *Brains On!*, American Public Media, December 26, 2017. https://www.brainson.org/shows/2017/12/26/the-nerve-electricity-in-our-bodies

Wilbur, Sarah. "Circadian rhythm pt. 2: More than human." Interview by *Brains On!*, American Public Media, March 13, 2018. https://www.brainson.org/shows/2018/03/13/circadian-rhythm-pt-2-beyond-human

PART 2: PLANTS

Braam, Janet. "Circadian rhythm pt. 2: More than human." Interview by *Brains On!*, American Public Media, March 13, 2018. https://www.brainson.org/shows/2018/03/13/circadian-rhythm-pt-2-beyond-human

"Common Cocklebur." University of California Integrated Pest Management. Accessed September 2019. http://ipm.ucanr.edu/PMG/WEEDS/common_cocklebur.html

"The Corpse Flower *Amorphophallus titanum*." New York Botanical Garden. Accessed September 2019. https://www.nybg.org/garden/the-corpse-flower-amorphophallus-titanum/

Finkle, Anita. "Q. Why are there so many acorns this year? Do they foretell a harsh winter?" New York Botanical Garden, April 2, 2018. http://libanswers.nybg.org/faq/222824

Garza, Ricky. "Carnivorous plants: How they lure, trap and digest." Interview by *Brains On!*, American Public Media, September 16, 2016. https://www.brainson.org/shows/2016/09/16/carnivorous-plants-how-they-lure-trap-and-digest

Grant, Richard. "Do Trees Talk to Each Other?" *Smithsonian Magazine*, March 2018. https://www.smithsonianmag.com/science-nature/the-whispering-trees-180968084/

"How a Tree Grows." Virginia Department of Forestry. Accessed September 2019. http://www.dof.virginia.gov/infopubs/_forest-facts/FF-How-A-Tree-Grows_pub.pdf

Kaufman, Rachel. "32,000-Year-Old Plant Brought Back to Life—Oldest Yet." *National Geographic*, February 23, 2012. https://www.nationalgeographic.com/news/2012/2/120221-oldest-seeds-regenerated-plants-science/#close

Klesius, Michael. "The Big Bloom—How Flowering Plants Changed the World." *National Geographic*. Accessed December 19, 2019. https://www.nationalgeographic.com/science/prehistoric-world/big-bloom/

Luby, Jim and Haskell, David George. "Tree, myself and I: All about our leafy green friends." Interview by *Brains On!*, American Public Media, July 16, 2019. https://www.brainson.org/shows/2019/07/16/tree-myself-and-i-all-about-our-leafy-green-friends

Musah, Rabi. "For crying out loud: All about tears." Interview by *Brains On!*, American Public Media, September 26, 2017. https://www.brainson.org/shows/2017/09/26/for-crying-out-loud-all-about-tears

Papia, Dominic. "Trees: From seed to shining seed." Interview by *Brains On!*, American Public Media, May 27, 2015. https://www.brainson.org/shows/2015/05/27/trees-from-seed-to-shining-seed

Pierre-Louis, Kendra. "We finally know how durian got so stinky." *Popular Science*, October 9, 2017. https://www.popsci.com/why-durian-smells-bad/

"Space Rose Pleases the Senses." NASA. Accessed December 19, 2019. https://spinoff.nasa.gov/spinoff2002/ch_1.html

Walker, Meredith Swett. "Harvester Ants Have a Taste for Exotic Seeds." *Entomology Today*, June 7, 2016. https://entomologytoday.org/2016/06/07/harvester-ants-have-a-taste-for-exotic-seeds/

Yong, Ed. "Plants Can Hear Animals Using Their Flowers." *The Atlantic*, January 10, 2019. https://www.theatlantic.com/science/archive/2019/01/plants-use-flowers-hear-buzz-animals/579964/

PART 3: HUMANS

Barrett, Deirdre and Martin Dresler. "Dreams: The science of the sleeping brain." Interview by *Brains On!*, American Public Media, December 11, 2018. https://www.brainson.org/shows/2018/12/11/dreams-science

Benditt, David and Jerry Vitek. "The nerve! Electricity in our bodies." Interview by *Brains On!*, American Public Media, December 26, 2017. https://www.brainson.org/shows/2017/12/26/the-nerve-electricity-in-our-bodies

Cleary, Anne. "Memory and deja vu." Interview by *Brains On!*, American Public Media, February 12, 2019. https://www.brainson.org/shows/2019/02/12/memory-and-deja-vu

Deo, Priya Nimish and Revati Deshmukh. "Oral microbiome: Unveiling the fundamentals." *Journal of Oral and Maxillofacial Pathology*, January–April 2019. https://www.ncbi.nlm.nih.gov/pmc/articles/PMC6503789/

Ennos, Roland. "Body bonanza: Yawns, hiccups, goosebumps and more!" Interview by *Brains On!*, American Public Media, October 25, 2016. https://www.brainson.org/shows/2016/10/25/body-bonanza-yawns-hiccups-goosebumps-and-more

Hattar, Samer and Martha Gillette. "The tick-tock of our circadian clock." Interview by *Brains On!*, American Public Media, February 27, 2018. https://www.brainson.org/shows/2018/02/27/the-tick-tock-of-our-circadian-clock

"Heart and circulatory system." Mayo Clinic. July 7, 2017. https://www.mayoclinic.org/diseases-conditions/heart-disease/multimedia/circulatory-system/vid-20084745

Hui, Alexandra. "Science under the microscope (Prove It: Part 2)." Interview by *Brains On!*, American Public Media, October 9, 2018. https://www.brainson.org/shows/2018/10/09/science-under-the-microscope-prove-it-part-2

Jeff, Janina. "Hey, where did you get those genes?!?" Interview by *Brains On!*, American Public Media, August 6, 2019. https://www.brainson.org/shows/2019/08/06/hey-where-did-you-get-those-genes

"Lungs." *National Geographic*. Accessed December 2019. https://www.nationalgeographic.com/science/health-and-human-body/human-body/lungs/

Marc, Robert. "Do we all see the same colors?" Interview by *Brains On!*, American Public Media, February 14, 2017. https://www.brainson.org/shows/2017/02/14/do-we-all-see-the-same-colors

Nelson, Kaz. "Sad: All about feelings, pt. 2." Interview by *Brains On!*, American Public Media, June 2019. https://www.brainson.org/shows/2019/06/11/sad-all-about-feelings-pt-2

Phillips, Kathryn. "Scratching the Fingernail's Surface." *Journal of Experimental Biology*, January 27, 2004. https://jeb.biologists.org/content/207/5/709.1

Romeo, Rachel. "What is dyslexia? And how do our brains read?" Interview by *Brains On!*, American Public Media, August 20, 2019. https://www.brainson.org/shows/2019/08/20/what-is-dyslexia-and-how-do-our-brains-read

Ross, Valerie. "Numbers: The Nervous System, From 268-MPH Signals to Trillions of Synapses." *Discover Magazine*, May 14, 2011. https://www.discovermagazine.com/health/numbers-the-nervous-system-from-268-mph-signals-to-trillions-of-synapses

Scudellari, Megan. "Cleaning up the hygiene hypothesis." *PNAS*, February 14, 2017. https://www.pnas.org/content/114/7/1433

Sheikh, Saira. "Allergy attack: How our bodies can overreact." Interview by *Brains On!*, American Public Media, July 28, 2016. https://www.brainson.org/shows/2016/07/28/allergy-attack-how-our-bodies-can-overreact

Skotko, Brian. "What is Down syndrome?" Interview by *Brains On!*, American Public Media, October 24, 2017. https://www.brainson.org/shows/2017/10/24/what-is-down-syndrome

Totten, Sanden. "New research explains why movies make us feel strong emotions." KPCC, December 4, 2014. https://www.scpr.org/news/2014/12/04/48457/what-watching-movies-can-tell-us-about-how-our-bra/

PART 4: MICRO-VERSE

Albala, Ken. "The science of baking." Interview by *Brains On!*, American Public Media, December 22, 2015. https://www.brainson.org/shows/2015/12/22/the-science-of-baking

"Are Viruses Alive?" Microbiology Society, May 10, 2016. https://microbiologysociety.org/publication/past-issues/what-is-life/article/are-viruses-alive-what-is-life.html

"The Beneficial Side of Pets' Bacterial Microbiome." University of Illinois College of Veterinary Medicine, February 11, 2018. https://vetmed.illinois.edu/pet_column/pets-microbiome/

Blackwell, M. "The fungi: 1, 2, 3…5.1 million species?" *American Journal of Botany*, March 2, 2011. https://www.ncbi.nlm.nih.gov/pubmed/21613136

"The Blob: Slime Molds." Utah State University Intermountain Herbarium. Accessed December 2019. https://herbarium.usu.edu/fun-with-fungi/slime-molds

Cassidy, Josh. "Meet The Mites That Live On Your Face." *KQED*, May 21, 2019. https://www.npr.org/sections/health-shots/2019/05/21/725087824/meet-the-mites-that-live-on-your-face

Fox-Skelly, Jasmin. "Tardigrades return from the dead." *BBC*, March 13, 2015. http://www.bbc.com/earth/story/20150313-the-toughest-animals-on-earth

"House Dust Mites." University of Kentucky. Accessed December 2019. https://entomology.ca.uky.edu/ef646

"Hydrothermal Vents." Woods Hole Oceanographic Institute. Accessed December 2019. https://www.whoi.edu/know-your-ocean/ocean-topics/seafloor-below/hydrothermal-vents/

Imster, Eleanor. "The largest land organism is…a fungus." *EarthSky*, December 6, 2017. https://earthsky.org/earth/largest-land-organism-honey-fungus

Knights, Dan. "Fart Smarts: Understanding the gas we pass." Interview by *Brains On!*, American Public Media, September 30, 2015. https://www.brainson.org/shows/2015/09/30/fart-smarts-understanding-the-gas-we-pass

Kong, Heidi. "Thinkin stinkin: Why we smell." Interview by *Brains On!*, American Public Media, April 30, 2019. https://www.brainson.org/shows/2019/04/30/thinkin-stinkin-why-we-smell

Witman, Sarah. "World's Biggest Oxygen Producers Living in Swirling Ocean Waters." *Eos*, September 13, 2017. https://eos.org/research-spotlights/worlds-biggest-oxygen-producers-living-in-swirling-ocean-waters

INDEX

ABOUT THE AUTHORS

Molly Bloom, Marc Sanchez, and Sanden Totten are the creators of American Public Media's *Brains On!* They became friends while working at Minnesota Public Radio and decided to make a show for kids, because kids are awesome. When they aren't making the show, they attend punk concerts, run half marathons, and make Swiss meringue buttercream (but you'll have to guess who does which activity). They invite you to visit them at brainson.org or @Brains_On.